Palgrave Macmillan's
Content and Context in Theological Ethics

Content and Context in Theological Ethics offers ethics done from theological and religious perspectives rooted in the particular contexts and lived experience of real people in history, in the present, and looking with hope toward the future. The series raises the contexts or cultures out of which an increasing number of scholars do their thinking and research regarding the influence of those contexts on the content of ethics and how that content has been applied historically, traditionally, and/or subversively by members of the context or community or culture under scrutiny or raised as paradigmatic or as a novel or passing fad. The series explores normative claims about right and wrong, human flourishing or failing, virtues and vices—the fundamental bases and questions of ethics—within the context, culture, or community identified and in correlation with norms inherited from or imposed by colonizing/dominant forces or ideologies while recognizing new voices and/or new understandings of theologically and/or religiously inspired concerns in response to knowledge uncovered by other disciplines, which impact ethical reflection on the content explored.

Series Editor:

MARY JO IOZZIO, active in the American Academy of Religion, Catholic Theological Society of America, Catholic Theological Ethicists in the World Church, Pax Christi USA, and the Society of Christian Ethics, she is a professor of moral theology at Barry University, Miami Shores, FL, and coeditor of the *Journal of the Society of Christian Ethics.*

Justice and Peace in a Renewed Caribbean: Contemporary Catholic Reflections
Edited by Anna Kasafi Perkins, Donald Chambers, and Jacqueline Porter

Theology in the Age of Global AIDS & HIV: Complicity and Possibility
By Cassie J. E. H. Trentaz

Constructing Solidarity for a Liberative Ethic: Anti-Racism, Action, and Justice
By Tammerie Day

Religious Ethics in a Time of Globalism: Shaping a Third Wave of Comparative Analysis
Edited by Elizabeth M. Bucar and Aaron Stalnaker

The Scandal of White Complicity and US Hyper-incarceration: A Nonviolent Spirituality of White Resistance
By Alex Mikulich, Laurie Cassidy, and Margaret Pfeil, with a foreword written by S. Helen Prejean CSJ

Spirituality in Dark Places: The Ethics of Solitary Confinement
By Derek S. Jeffreys

Narratives and Jewish Bioethics
By Jonathan K. Crane

NARRATIVES AND JEWISH BIOETHICS

Jonathan K. Crane

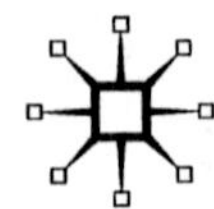

NARRATIVES AND JEWISH BIOETHICS
Copyright © Jonathan K. Crane, 2013.

All rights reserved.

First published in 2013 by
PALGRAVE MACMILLAN®
in the United States—a division of St. Martin's Press LLC,
175 Fifth Avenue, New York, NY 10010.

Where this book is distributed in the UK, Europe and the rest of the World, this is by Palgrave Macmillan, a division of Macmillan Publishers Limited, registered in England, company number 785998, of Houndmills, Basingstoke, Hampshire RG21 6XS.

Palgrave Macmillan is the global academic imprint of the above companies and has companies and representatives throughout the world.

Palgrave® and Macmillan® are registered trademarks in the United States, the United Kingdom, Europe and other countries.

ISBN: 978–1–137–02616–3

Library of Congress Cataloging-in-Publication Data

Crane, Jonathan K. (Jonathan Kadane)
Narratives and Jewish bioethics / Jonathan K. Crane.
pages ; cm. — (Content and context in theological ethics)
Includes bibliographical references and index.
ISBN 978–1–137–02616–3 (hardcover : alk. paper)
1. Bioethics—Religious aspects—Judaism. 2. Bioethics in literature. 3. Jewish literature—History and criticism. I. Title.
BM538.B56C73 2013
296.3′691—dc23 2012038794

A catalogue record of the book is available from the British Library.

Design by Integra Software Services

First edition: March 2013

10 9 8 7 6 5 4 3 2 1

For Lindy
My storyteller without compare

Contents

Series Editor's Preface

Content and Context in Theological Ethics, as a new series in the Palgrave Macmillan titles in religion, offers a fresh look at the millennia-old tradition of ethics engaging religions, their scriptures and revered texts, and their theological reflections on what matters and why. The series is first and foremost focused on ethics, done from theological and religious perspectives, and rooted in the particular contexts and lived experience of real people in history, in the present, and hoped for in the future. While engaged by diverse contexts, themes emerging in the series span the gamut of research in ethics that provoke theological and/or religious concerns; for example as this text demonstrates, ethical reflection on the narratives that form the bases of bioethics engaged in the traditions and thought of Jewish scholars from antiquity to contemporary authorities. Since contemporary work in ethics is increasingly context driven and characterized by diversity, this series brings contextual theological and religious ethics to bear on the content explored.

Narratives and Jewish Bioethics by Jonathan Crane brings new insight into and practical suggestions for thinking ethically about how the centuries' old tradition of narratives as teaching devices informs contemporary bioethics discourse. As with other subjects engaged in this series, inherited traditions are measured through detailed analysis of the concrete/context-laden lived experience of the people the traditions claim as their own and critical reflection on what was the past for them and/or what could be the future. The series provides scholars with books of interest on a broad range of subjects in ethics identified with a particular community whose voice and experience are underrepresented in ethics, theology, religious studies, and related disciplines. In this text, Jonathan Crane retrieves a classic story told and retold, parsed and reparsed through time. Crane considers how one story in particular, the burning death of Rabbi Chananya ben Teradyon, influences Jewish ethical norms and how those norms may influence interpretation of the narrative.

I am privileged to include in the series' first years of publication a text so thoroughly steeped in the Jewish bioethical tradition. As in these times the debate over euthanasia looms large, on account of the almost two-decade permissibility of active voluntary euthanasia practices in The Netherlands and the three state provisions regarding Physician-Assisted Suicide, *Narratives and Jewish Bioethics* raises the debate a notch higher and makes its substance deeper. What Crane brings to the debate is a narrative that has been used to advocate and reject arguments for all kinds of euthanasia. What readers will find is how this and other narratives may be understood by examining the complex relationship between the stories and the law. That relationship is key to determining the context that will illumine the content of the tradition.

As I write, the United States prepares for yet another presidential election in which questions regarding "death panels" and the sanctity of human life continue to press the two major political parties and the US public. Euthanasia is one part of those questions and an important one for those who recognize the vulnerability of life and the requisite defense of justice for those who find themselves wondering how to proceed. While this text does not answer the question, it does provide the intellectual nuance necessary for any serious consideration of what constitutes—in both bioethical and Jewish senses—a good death.

Readers, welcome to the series!

Mary Jo Iozzio
Series Editor

Acknowledgments

Every story begins with a seed, and every time one returns to that seed to nurture it, one discovers within it new things. About this, the Talmud teaches: R. Chiya b. Abba in the name of R. Yohanan taught: What does the scripture mean, *Whoever keeps the fig tree shall eat thereof* (Proverbs 27:18)? Why are the words of Torah likened to a fig? Every time a person attends to a fig tree, he or she will find figs therein. So too the words of Torah: every time a person ponders them, he or she will find pleasing wisdom (BT *Eiruvin* 54a–b). The seed for this story emerged from within my dissertation studies with Robert Gibbs and David Novak at the University of Toronto. It took root at the Society of Jewish Ethics conference in New Orleans in January 2010, where I presented a seedling of this unfurling narrative. To these great teachers and those wonderful colleagues, I express my humble gratitude. I am also grateful to Mary Jo Iozzio, then the coeditor of the *Journal of the Society of Christian Ethics*, for expressing profound enthusiasm for this project and inviting me to expand it for this series at Palgrave Macmillan. Many thanks are due to the anonymous reviewers for their initial pruning and subsequent weeding.

As this juicy story began to ripen, I gratefully received advice and wisdom from fellow tenders of the Judaic textual tradition. Thanks go to Barry Wimpfheimer, Elliot Dorff, Julia Watts Belser, Aaron Gross, Ed Elkin, Louis Newman, William Cutter, and Toby Schonfeld. Their insights helped keep wayward limbs from twisting away and toppling the overarching narrative. The great staff and faculty at the Center for Ethics at Emory University have been exemplary colleagues, always offering animated support whenever I ventured out to stretch my legs. I also thank the baristas at Dancing Goats in Decatur, Georgia, for quenching my thirst as I bent on this task. I appreciate Matthew LaGrone's careful reading and development of the index.

My sons, Nadav and Amitai, came into this world to see this project blossom. Their never-diminishing enthusiasm to hear yet another book further prompted me to finish this one. Their curiosity and optimism kept my own fresh. But it is to my wife, Lindy Miller, I devote this volume, for it is she who has constantly tended my doubts, doused me with both realism and reassurance, and pointed to new possibilities.

About that fig tree, Abraham Joshua Heschel wrote in *God in Search of Man*, "The Bible is a seed, God is the sun, but we are the soil. Every generation is expected to bring forth new understanding and new realization." To those of you reading this story of stories and find new fruit therein, thank you, for through you the orchard of Jewish bioethics continues to expand and bloom.

Abbreviations

BT	Babylonian Talmud (Bavli)
JT	Jerusalem Talmud (Yerushalmi)
M	Mishnah
MT	Mishneh Torah
SA	Shulchan Aruch
T	Tosefta

[M]an is in his actions and practice, as well as in his fictions,
essentially a story-telling animal. He is not essentially,
but becomes through his history, a teller of stories that
aspire to truth.
But the key question for men is not about their own authorship;
I can only answer the question
"What am I to do?"
if I can answer the prior question
"Of what story or stories do I find myself a part?"

Alistair MacIntyre

But equally, historians need to treat a participant's
own explanation of events with a certain skepticism.
It is often the statement made with an eye to the future
that is the most suspect.

Julian Barnes

Chapter 1

Genesis of Jewish Bioethics

Stories are lived before they are told—except in the case of fiction.[1]

Introduction

What *can* be done for and to the dying certainly differs from what *should* be done, yet discerning between them is difficult. The dual move of deciphering the realm of the possible in regard to such care and deciding the narrower category of the preferable is inherently an ethical endeavor. It involves as much imagination as wisdom, since advocating just any or all kinds of activities would be dangerous. Hence, care must be taken when thinking about the care to be given.

Creating norms for such care occurs "in a field of pain and death," as the legal philosopher Robert Cover once said in regard to legal interpretation by judges.[2] Sometimes lethal by design and more often fatal only inadvertently, creating norms for end-of-life care is challenging because it involves real people (e.g., dying patients, families, professional care providers, clergy, and community in general). Furthermore, those creating the norms are persons who are themselves caught up in the throes of life, overflowing as it is with political and theological currents, familial concerns, and technological advances (the benefits and demerits of which have yet to be fully appreciated). For all these reasons, it behooves those of us who take guidance from scholars, professional care providers, clergy, and others who spend time and energy thinking seriously about how we can and should care for the dying to appreciate both *how* and *why* they make the arguments that they do. This is because there is a story about bioethical norms, a story that is at once historical, theological, legal, literary, and, ultimately, existential.

Tracing this story is less than straightforward, however. Details and digressions may distract one's attention from the larger narrative of how and why bioethical norms regarding the dying have come about. Lest the maze of law or the amazing prowess of modern medicine entice and entrap, discipline is required to shorn this story of excesses and keep it on track. An unwavering focus promises to assist the helpers and the helped. To this end, the story told here hones in on a particular classical narrative and how it pervades modern Jewish bioethical deliberation on care at the end of life. It will be shown that this particular story is no simple tale. Its complexity, ambivalence, and ambiguity pose as problems, challenges, and opportunities for contemporary bioethicists. How they read and wrestle with *that* story is part of what makes *this* story so fascinating. What follows—the first of its kind—is but one version of what could be said about narratives and modern Jewish bioethics.[3] More can and should be said on this subject, to be sure, but this should not hinder us from telling at least a first draft of this important and evolving story that has life-and-death implications and applications.

Some readers may find this story incendiary because it questions the oft-unquestioned reasoning of certain modern norm-creators, or because it scrutinizes the presumed authority of certain classic texts, or because it investigates the very relationship between narratives and norms—a relationship that hitherto has received scant attention in modern Jewish bioethics. My intention is not to burn bridges, sources, or conversations. Rather, my goal is to spark and invigorate interest in the complex interrelationship between narratives and norms, especially in regard to the ever-smoldering issues surrounding end-of-life care.

Story Matters

The great modern moral philosopher Alasdair MacIntyre realized he could truly appreciate his subject matter (the ancient Greeks and Aristotle with his virtues in particular) only if he paid attention to the stories they told—stories of and for themselves, stories of and for the world. "There is no way to give us an understanding of any society, including our own, except through the stock of stories which constitute its initial dramatic resources. Mythology, in its original sense, is at the heart of things."[4] If mythology—the study of legends, lore, and stories—is at the heart of all things philosophical, it is surely no less true in regard to matters that are less ethereal and abstract but more earthy and practical.

And if mythology's importance to societies at large is undeniable, the relevance of stories to individuals comprising any society cannot rightfully be suppressed or sidelined. Whereas it is commonly understood that people make stories, the inverse is also true: stories make people. Stanley Hauerwas, a renowned Christian theological ethicist and perhaps the greatest proponent of appreciating theology through narrative, holds that "To be moral persons is to allow stories to be told through us so that our manifold activities gain a coherence that allows us to claim them as our own. Our experience itself, if it is to be coherent, is but an incipient story."[5] Though my story may be incipient, its beginning does not begin with me, however. This is as true about this project as it is about each and every person.

Each person emerges into already ongoing stories, and they run their courses in, through, and around us—whether we like it or not. These infinitely complex interweaving stories in which we are inextricably embedded *are* history: they undeniably precede us and surely will outlive us. As MacIntyre says, "What I am, therefore, is in key part what I inherit, a specific past that is present to some degree in my present. I find myself part of a history and that is generally to say, whether I like it or not, whether I recognize it or not, one of the bearers of a tradition."[6] However much or little we like the stories in which we find ourselves—and to which we inexorably contribute—they constitute who we are, what our life's projects are, and what it all means to us. This notion of inheritance continues in Hauerwas, who insists that "The metaphors and stories we use to organize our life plan are inherited from our culture and our particular biographical situation," as if to say, as would the French existential philosopher Emmanuel Levinas, that we come into existence already bequeathed and endowed with—and indebted, really, to others for—the stories, meanings, and obligations that constitute our life's context and shape our prospects.[7]

We may be late to the ongoing saga that is our social world but we are not impotent to contribute thereto. Indeed, ours are unique narrative existences; we are subjects unique unto ourselves inscribing our own stories. Hence MacIntyre's twofold notion of a "narrative concept of self" begins with subjecthood: "I am what I may justifiably be taken by others to be in the course of living out a story that runs from my birth to my death; I am the *subject* of a history that is my own and no one else's, that has its own peculiar meaning."[8] Insofar as I am uniquely situated in this peculiar matrix of relationships and responsibilities, no one else can or even could take my place, much less usurp my story. To this Levinas would certainly agree, as well as to the fact that I cannot be extricated from my situation, my narrative embeddedness. The idiosyncrasy of my story is inalienable: no one can excise me from it nor assume it for himself or herself.

Since my story *is* mine, others can, should, and do ask me to explain myself—to understand me and my story, to make it intelligible to them (and to me!). In their asking me to articulate why I did what I did or why I do what I do, they seek an accounting from me—and this is the second aspect of MacIntyre's "narrative concept of self." They want me to recount my reasons, reveal my motivations. Of course this exchange does not always occur verbally. The raised eyebrow, as much as the spoken "why?", prompt me to explain myself. Though they cue my story, I am the one caught up in it. In this way, through others I am beholden to my story, to myself. Yet my story is not infallible or perfect, try as I might. For better or for worse, however hard we try to express ourselves truthfully, our self-narratives are fallible because our memories of ourselves are vulnerable to delusions, elisions, or evasions.[9]

Being accountable for myself is more than a reflexive responsibility, however. It is also transitive—for I, too, am bound up in others' stories and perforce seek from them their own accounting. Others are also beholden to me, according to MacIntyre:

> I am not only accountable, I am one who can always ask others for an account, who can put others to the question. I am part of their story, as they are part of mine. The narrative of any one life is part of an interlocking set of narratives. Moreover this asking for and giving of accounts itself plays an important part in constituting narratives. Asking you what you did and why, saying what I did and why, pondering the differences between your account of what I did and my account of what I did, and *vice versa,* these are essential constituents of all but the very simplest and barest of narratives. . . . Without that same accountability narratives would lack that continuity required to make both them and the actions that constitute them intelligible.[10]

I would push MacIntyre here: I am not just one who *can* ask from others an accounting, I am one who *cannot* do otherwise. Just as everyone else constantly requests me to explain myself, my very existence is a series of questions seeking from others why they do what they do. I cannot but ask because I am, Levinas insists, responsible for your and everyone else's responsibility.[11] Each person, Levinas observes, comes into the world scene already indebted and excessively endowed with duties. Each person navigates—narrates—the repayment of this debt and the fulfillment of these never-ending obligations by being accountable to and extracting from others their own accounts. Since the pieces of the social scene ever shift, no account is stable for long. New questions arise, sparking a fresh round of critical self-reflection in each and every person. Life is thus questionable, an ongoing search for certainty in an ever-adjusting exchange of perspectives and explanations. It is this iterative process of offering and receiving accounts that constitutes life's intelligibility.

This give-and-take between people is akin to Lenn Goodman's notion of chimneying. In his Gifford Lectures on the Levitical command to "*love thy neighbor as thyself*" (19:18), Goodman suggests the image of chimneying climbers "who push off opposing rock faces as they work their way upward in a narrow defile" to speak of "an ongoing dialectic between ethics and religion, as our insights about value, including moral value, inform and are informed by our ideas about the divine."[12] The relation between ethics and religion—that is, between extant morality and revealed morality—is dialectical precisely because "we bring our moral notions, suasions, customs, instincts, attitudes and intuitions to the Law [revelation], and they enter into dialogue with what we read, informing our hermeneutic, as scripture itself and the conception of God encountered in scripture, inform them in turn."[13] Neither edifice—ethics or religion—is unmoving, foundational, or non-question-begging. Both shift and adjust as people chimney between ideas of God and ideas of other values. Our hermeneutics—our questions of texts and of people—continuously form and are informed by the texts and people we encounter.

We ever shimmy along our unique paths between each other as much as we do between our ideas of ethics and our notions of revealed morality. Goodman continues with this theme:

> When the Torah offers *Love thy neighbor as thyself* to warrant the duty of reproof, the application shapes the generality. For morals, like the sciences, can work inductively. And induction is just as able to show off the middle terms on which an inference hinges as it is to vault from cases to the universal rule that links them or rappel back again to further cases. The middle term here, connecting the particular to the universal, is our common, yet unique personhood, whose boundaries are illuminated morally by the bans that hedge it about.[14]

Insofar as personhood fluctuates by embodying the universal in the practical, it is more verb than noun, more movement than stasis.

On this point we bound to the French philosopher Paul Ricoeur, who echoes these themes, though his is a response to MacIntyre's theory of narrative unity and speaks not of chimneying but of a twofold movement between universals and particulars:

> Here I shall attempt to bring to light the simple fact that the practical field is not constituted from the ground up, starting from the simplest and moving to more elaborate constructions; rather it is formed in accordance with a twofold movement of ascending complexification starting from basic actions and from practices, and of descending specification starting from the vague and mobile horizon of ideals and projects in light of which a human life apprehends itself in its oneness. In this sense, what MacIntyre calls "the narrative unity of a life" not only results from the summing up of practices in a globalizing form but is governed equally by a life project, however uncertain and mobile it may be, and by fragmentary practices, which have their own unity, life plans constituting the intermediary zone of exchange between the undetermined character of guiding ideals and the determinate nature of practices.[15]

Goodman's "middle term" and Ricoeur's "intermediary zone" are the stories of which we are speaking: they are our unique lives, inextricably bound up with each other as they are with life's minutiae and immensity. We are forever bouncing between the imperatives of the immediate moment and eternal aspirations. Neither face—the face of the other facing me just now nor the face of the infinitely distant revelation or redemption, nor my own face for that matter—is absolute or firm. Their clarity and trustworthiness emerges through iterative encounters and novel questions.

Asking new questions is itself a kind of quest. It is a sort of journey through which the *telos* or goal becomes increasingly intelligible. On this point we bound between Ricoeur and Hauerwas. The former says:

> It is in the course of the quest and only through encountering and coping with the various particular harms, dangers, temptations and distractions which provide any quest with its episodes and incidents that the goal of the quest is finally to be understood. A quest is always an education both as to the character of that which is sought and in self-knowledge.[16]

Hauerwas sees it thus:

> [C]ertain prohibitions of a community are such that to violate them means that one is no longer leading one's life in terms of the narrative that forms that community's understanding of its basic purpose. For the *telos* in fact is a

narrative, and the good is not so much a clearly defined "end" as it is a sense of the journey on which that community finds itself.[17]

The transgressions of which Hauerwas speaks reflect Ricoeur's harms and dangers. These "what we ought *not* do" are necessary boundaries that become increasingly intelligible in the very process of conversing, of asking, of holding each other accountable. We come to understand who we are by clarifying to ourselves and others what it is that we pursue and what it is that we should not do.

But of these, it is that which is proscribed that is the most telling, or revealing, of a person's or community's conviction and character. Thus Hauerwas holds, "A community's moral prohibitions, therefore, are not so much 'derived' from basic principles as they are from the way the community discovers what in fact its commitments entail."[18] Note that the *way* a community discovers its commitments requires focusing on how its prohibitions are articulated in the first place. Are they promulgated by some revelation? Perhaps they are imposed by fiat. Or are they deliberated through some collective process? Reflexively interrogating both *what* a community prohibits and *how* it goes about identifying the realm of the unacceptable uncovers—that is, discovers—both the *character of that which is sought and the self-knowledge* of which Ricoeur speaks. The quest for clarity for what we should not do (and by extension, what we should do) narrates the rules that guide our daily lives as well as our larger personal and collective projects.

Stories matter no less for individuals than they do for communities. For through stories, persons and societies alike gain clarity on what matters to them and why. The importance of stories is not lost on Hauerwas when he insists that "We neither are nor should we be formed primarily by the publicly defensible rules we hold, but by the stories and metaphors through which we learn to intend the variety of our experience. Metaphors and stories suggest how we should see and describe the world—that is how we should 'look-on' our selves, others, and the world—in ways that rules taken in themselves do not."[19] This is not to say that rules like laws, norms, or principles are useless markers of what a society or a person holds dear. They are this, to be sure. But not all laws, norms, and principles are enacted; not all shape human behavior and experience, nor do they necessarily inform how we see the world at large. Many laws, for example, go unheeded or unenforced. That said, *how* those laws, rules, and norms came into being in the first place is itself a story, a narrative about what matters to a community and why.

CONTEXT MATTERS

This project is itself a story. It recounts the role of narratives in contemporary Jewish bioethics. In so doing it simultaneously encounters other modern Jewish bioethicists and asks of them to account for how and why they do what they do. To keep this story going, to keep it surprising, it hones in on a particular story, a narrative to which Jewish bioethicists repeatedly turn when constructing laws, rules, and norms. So, to be precise, this is a story of a story.

But before we can turn the page to that particular modern story, we must appreciate its context. This story is indebted to an old, complex story that traverses millennia, continents, worldviews, languages, technologies, and even theologies. It emerges into a scene whose own provenance is lost in the mists of ancient myths.

With no pretenses of recapitulating the totality Jewish medical history, suffice it to say that Jews and Judaism have long been concerned with physical well-being. Such concern emerged in a context already imbued with significant, even sophisticated, biomedical knowledge. Exposed to nearby conceptualizations of health, illness, and death in ancient Egyptian and Mesopotamian cultures, early Judaism developed its own notions of well-being, its approximations, and its converse.[20] Whereas those other peoples worked within polytheistic worldviews in which illness was caused by a divinity or a demon or by magic or perhaps by nature, the Jewish monotheistic paradigm functioned with the background theory that God is the source of all ailments—and healing.[21] This assumption, however, did not prevent Jews from appreciating the fact that human suffering and eventual death often result from natural and human causes. For example, Rabbi (hereafter, R.) Chanina opines that all ailments are from God except for colds, and colds are a major contributor to mortality.[22] R. Acha asserts that it depends upon each individual whether he or she becomes ill—suggesting that human behavior (religiously transgressive or not) is the prime source of suffering.[23]

Such notions point to a significant thread weaving throughout this story—the thread of theodicy, the relationship between divine justice and earthly anguish. Specifically, if Jews assume that God is good (itself a contentious theological claim, to be sure, but one that certainly has grounding in the Judaic textual tradition), how can God allow humans to suffer and die? Though the rabbis of old assumed that no human exists without torments, the question still stands.[24] If it is not ultimately random, what might be the meaning or purpose of human pain and mortality? Such questions vexed Jews throughout the eras with nearly every layer of the Judaic textual tradition countenancing multiple theodicies.

The varieties of Judaic theodicies can divide into four major camps, three of which are arranged temporally.[25] Many *past-oriented theodicies* understand human suffering in terms of some kind of punishment for wrongdoing.[26] Sin is often considered to be the primogenitor for all kinds of physical and spiritual suffering. Summarizing Ezekiel's notion of individual responsibility (18:20) and the Psalmist's insistence that God punishes waywardness (89:33), R. Ammi said: "There is no death without sin, no suffering without transgression."[27] Elimination of sin would, in theory, bring about relief from suffering, or at least keep ailments at bay.[28] Cure could be found in *teshuvah,* repentance.

Present-oriented theodicies look upon suffering more as a kind of spiritual exercise or as an opportunity for spiritual enhancement. Job famously understood his suffering as a test, trial, or exercise of his fidelity to God (2:10). R. Judah HaNasi once quipped that suffering is precious, echoing the idea that sufferers should rejoice since through their pains they achieve divine forgiveness, that is, they come ever closer to God.[29] And while he was being executed for teaching Torah publicly, R. Akiva explained jubilantly that only now could he fully appreciate the meaning of the Shema prayer.[30] Ailments and existential suffering are no sources of shame but opportunities for improvement and deeper spirituality.

Future-oriented theodicies also see suffering as an opportunity but understand it more as pedagogy or therapy. Nahum of Gamzo, for example, suffered greatly from quadriplegia and full-body sores. He claims that he brought this condition upon himself for not attending quickly enough to another's hungry impoverishment. Others cry out, "Woe that we should see you in such straits!" Nahum replies, "Woe to me if you had not seen me in such straits."[31] In this way, Nahum leverages his suffering into a pedagogical tool to teach others that they too have the capacity to do well by others and for themselves.[32] And R. Yannai would wear his *tefillin* (prayer phylacteries) all day long for three days to teach that ailments themselves cleanse the body of sin; that is, suffering itself is therapeutic.[33]

The last kind—*mystery theodicies*—emphasize that the provenance and meaning of human suffering is ultimately unknowable. Some, like Moses Maimonides, the great twelfth-century philosopher, legist, and physician, hold that humans can never fully understand the ways and whys of God's will manifesting in the world, especially when it comes to human suffering.[34] Sometimes human suffering is best understood as instantiating God's affections: they are but a symptom of God's love.[35] Taking this notion further and beyond this world altogether, some hold that the sufferer in the here and now will reap rewards in the World to Come.[36]

It would be wrongheaded to assume theodicy does not come into play in contemporary Jewish bioethical deliberation, especially in regard to care at the end of life. Indeed, as this story unfolds, the evidence that theodicy plays a significant role will be undeniable. It appears in all its complexity in the central classical story as much as it does in the ways modern Jews read that story. And insofar as theodicy is but a small slice of theology, this suggests that bioethical deliberation occurs within the milieu of and with the tools of theology. Again, this suggests that Jewish bioethicists chimney between suffering and meaning, between the real and the religious, between experience and aspiration, ever in search of greater understanding of each.

Another strand of Jewish thinking weaves throughout this discussion of how and why humans suffer: the string of human agency. Once we acknowledge human suffering's physical urgency and theological significance, the question now turns to what we humans can and should do about pain, morbidity, and mortality. Even though God is understood as the ultimate source of illness and well-being, humans are nonetheless tasked to be God's partners in bringing about health.[37] More than a mere permission, this task is a duty. According to the Talmud, humans are obliged to help heal.[38] This existential and creative responsibility inspired many sages to meditate on and spell out the ways humans can and should protect human health, prolong human life, and usher humans on toward eternity.[39] Such concerted efforts over the centuries for a singular purpose does not mean that there is a central text to which all Jews would turn when confronting biomedical problems. With only a few medieval exceptions, no classic Judaic source devoted exclusively either to medicine or to bioethics exists—a fact frequently lamented by scholars of the history of Judaism and medicine.[40]

Two of the exceptions are the early medieval *Book of Asaph,* and some of Maimonides's essays. The former was composed by Asaph ben Berakhiah before the sixth-century CE in Israel or Syria (this early date is surmised from the absence of references to Arab or Muslim sources). It is a compilation of medical knowledge drawn from a variety of cultures: ancient Hebrew of course, as well as Babylonian, Egyptian, Persian, and Indian. Inspired by Galen of Pergamon, the famous second-century CE Roman physician, and Hippocrates, the fourth-century BCE Greek physician, Asaph understood the body to be composed of humors that, when imbalanced, brought about illnesses. He details medicinal plants, discusses recipes for pharmacological concoctions, teaches ways to identify and diagnose diseases, and, more interestingly, offers strategies to prevent ill-health (e.g., exercise, restricted diet, hygienic habits).[41]

The more famous exception is the work of Maimonides. He was such a great physician in Fostat, Egypt, that the Vizier al-Qadi al-Fadil al-Baysani introduced him to the new sultan, Salah al-Din (Saladin), somewhere around 1171 CE.[42] Also inspired by Hippocrates and especially by Galen's physiology, Maimonides nonetheless drew heavily from Muslim and other Arab medical practices. In fact, his various medical treatises were written in Arabic. As was his wont, he promoted care based on the integration of three sources of knowledge: reason, experience, and tradition. This is because analyzing, attending to, and interpreting human suffering perforce requires intertwining philosophy, science, and religion. Like Asaph, Maimonides delved into particular conditions and diseases, and also prescribed ways to prevent ill-health from arising in the first place through dietetics, exercise, and hygiene. Still, he did not produce a single tome in which all his medical knowledge could be found, nor did he render absolute rulings on what to do with complicated cases.

CONTENT MATTERS

Such medically interesting sources notwithstanding, the best that modern bioethicists can do is to plumb the massive Judaic textual tradition to find appropriate tidbits here and there. Testifying to this and perhaps offering the best model of it is Julius Preuss's 1911 *mangum opus:* the *Biblisch-Talmudische Medizin* (Biblical and Talmudic Medicine).[43] He scours the Judaic textual tradition for any bit that might relate to medicine. His work far surpassed earlier nineteenth century efforts to catalogue ancient Jewish medicine because he himself "was a first class physician, who made the history of medicine his life's study, and [he] was a thorough semitic philologist."[44] With both medical expertise and mental prowess, Preuss (1861–1913) organized the book in light of the nineteenth-century optimism of humankind's abilities to manipulate the world. For this reason, it opens with a description of physicians and other professional care providers. He then turns to the human body itself, detailing its anatomy and physiology—as it was understood in antiquity, and afterwards goes on to define the patient and its corollary: disease. The rest of the book adumbrates various sicknesses, conditions, and disorders, and traces how the ancients treated them through pharmacological intervention, behavior alterations, even surgery. In the last few chapters he reviews classical views on sex ethics and dietetics in its fullest sense—meaning a regimen of daily health. As such, this work provided unparalleled access to ancient Jewish medicine.

What Preuss did in the first part of the twentieth century for European Jewry, Immanuel Jakobovits did for English-speaking Jewry 50 years later. Having been Chief Rabbi of Ireland, Jakobovits (1921–1999), then a rabbi of the Fifth Avenue Synagogue in New York, published in 1959 what was his PhD dissertation, a work he admits was prompted by advances in Catholic medical ethics.[45] This book, *Jewish Medical Ethics,* took Preuss's model further. Like his predecessor, Jakobovits organized his book into clear chapters that made it easy to peruse. He also swept through the classic Judaic sources, weaving together bits from here and there, and threaded them with the works of ancient scholars and physicians from other cultures (hence the subtitle: "a comparative and historical study of the Jewish religious attitude to medicine and its practice"). But whereas Preuss's corpus begins with the human body (physician, physiology, patient), these take second stage in Jakobovits's. God pervades the beginning here: the first chapter opens with a discussion of human healing in light of divine providence. This sets the backdrop for the next seven chapters that deal with laws, rules, and regulations pertaining to the duty to heal, how, when, and by whom. Only in the ninth chapter is the sick patient introduced and discussed as such. The book ends with five chapters exploring the role and responsibility of the physician. Here the body takes second fiddle to law; the physician trails the patient.

The writing within the chapters complements this macro-structure that situates law before all else. Where Preuss merely reported what exists in the classic sources, Jakobovits chimneys between those (and other) sources and modern science to construct assimilative and summative arguments. Preuss describes Jewish biomedical practice; Jakobovits prescribes. En route he articulates definitive positions on practical biomedical issues. The very notion that there can and should be definitive conclusions on pressing biomedical issues inspired subsequent generations of bioethicists to scour the tradition in search of precedents and analogous legal cases. This frenetic search for clear conclusions perforce meant eschewing other classic texts, such as non-legal narratives, liturgy, history, and even science itself to a degree. Law came to dominate the field of modern Jewish bioethics.

Or at least this became the leading model within Orthodox Jewish bioethical discourse, especially in its English-medium camp. Here contributors—such as doctors Fred Rosner, Avraham Steinberg, Mordechai Halperin—often drew from their clinical and medical experiences. Working alongside them were their counterparts whose primary training was in rabbinics, including such scholars as Moshe Tendler, Moshe Feinstein, J. David Bleich, and Eliezer Waldenberg. That law serves as their primary lens through which they examine and deliberate perplexing bioethical and morally troublesome issues derives from the overriding Orthodox assumption that no distinction between Jewish law (*halakhah*) and morality exists. Steinberg summarizes this most clearly:

> In Judaism, there is no basic difference between laws and regulations and morals and ethics because both are integral parts of the Torah and their validity flows from the power of the Torah and the Divine revelation.[46]

Though this assumption is controversial and has been challenged even by Orthodox scholars, it nonetheless remains the overarching *modus operandi* for these bioethicists.[47] Precedents thus inform their arguments, as do principles. Identifying such appropriate foundations then becomes the paramount task for modern bioethicists. In the view of the most eloquent scholarly observer of modern Jewish ethics, Louis Newman, Orthodox bioethicists *discover* norms in the Judaic textual tradition—for everything can be found therein if only one looks hard and well enough—an observation Bleich explicitly endorses.[48] Such procedural patterns reinforce the traditional stance that the rightful authority figure and proper decision-maker in regard to bodily issues like healthcare and end-of-life matters is none other than the rabbi. Bleich advocates this clearly when he says physicians must comport to Jewish law, even in the face of competing science and societal mores, and families, too, must step aside so that rabbis can do their job of assessing the status of a patient and deciding what sort of treatment is legally permissible and required.[49] A doctor-rabbi should hearken more to his traditional training than his scientific knowledge.

Not all Orthodox bioethicists toe this line, however. Some, like Noam Zohar, take inspiration from Moshe Feinstein (1895–1986), often revered as the most authoritative halakhic decisor of the twentieth century, who broke from absolute deference to the institution of the rabbi. Feinstein, whose Hebrew works are only slowly being translated into English for a broader hearing, understood the individual patient—or the patient's healthcare proxy if the patient is incapacitated—to be the proper authority to make healthcare decisions. Feinstein makes this explicit when he rules that a physician is not required to cure a patient if it would only secure enduring suffering for the patient, a rule to which his son-in-law Moshe Tendler added, "it is a decision which the patient must make."[50] Feinstein also breaks from the strict legal formalism so ensconced among the Orthodoxy. To illustrate, he invokes the story of Rabbi Judah HaNasi's final moments of suffering greatly from a gastrointestinal disease to justify his ruling protecting patients from receiving medications if it would only prolong their suffering.[51] More will be said about the turn to narratives in the next chapter. Despite such innovations, Feinstein remained a critical figure in modern Orthodox Jewish bioethics inasmuch as contemporaries and subsequent generations continuously turn to his decisions for normative guidance.

Certainly Jewish bioethicists emerged outside the Orthodox camp. Elliot Dorff, Aaron Mackler, David Feldman, Avram Reisner, and Leonard Sharzer, among many others, have led bioethical deliberation within the Conservative movement. With a few exceptions, they are rabbis who took a strong interest in medical issues. And although they work within Conservative's impulse to be beholden to law, they nonetheless deferr less to a particular rabbi than to the community, as instantiated by the movement's Committee on Jewish Law and Standards. Like their Orthodox counterparts, they also plumb the textual tradition in search of legal strictures and analogous cases. But they are more willing to situate those sources within their historical contexts, an admittance that opens up their moral arguments to additional influences, such as from science and modern moral sensibilities.[52]

Leaders of Jewish bioethical discourse among the Reform community have contributed to volumes produced by the Freehof Institute's Progressive Halakhah series. A few of these rabbinic scholars include Walter Jacobs, Moshe Zemer, Mark Washofsky, Leonard Kravitz, and William Cutter. The prominent assumptions here are that the individual is the rightful authority to make normative decisions and the best authors and rabbis can do is to somehow persuade them using creative arguments intertwining Judaic and contemporary sources. A recent volume of the *Central Conference of American Rabbis Journal* was devoted to a symposium's discussion of Judaism, Health and Healing.[53] The essays demonstrate the variety of ways the individual is championed as the proper decision-maker since it is the individual patient who experiences a particular condition—and not a rabbi or physician or other care provider. Thus this volume illustrates one of the challenges facing non-halakhic bioethics inasmuch as modern notions of autonomy militate against directing or constraining the ways people should treat their idiosyncratic ailment experiences.

Whereas the above bioethicists work explicitly from within Judaism, other bioethicists turn to the Judaic textual tradition to buttress their otherwise philosophical and medical work. For example, Laurie Zoloth, Baruch Brody, and Benjamin Freedman are first and foremost professional bioethicists with medical training, yet they deploy Judaic resources and reasoning to illustrate and augment their arguments. In Newman's view all these non-Orthodox styles of modern Jewish ethical deliberation *create* norms more than they discover them from within the Judaic textual tradition. This is not to say that Orthodox bioethics and non-Orthodox bioethics are distinct genres. That conclusion would sever the field as much as it would kill the efforts of many to engender broader conversations across these camps. It would also assume that these differences are of a kind; they are not: they are merely differences of degree. Jewish bioethics is indeed a large, diverse yet identifiable discourse and field.

Despite such differences, all contributors to modern Jewish bioethics deploy hermeneutics, or reading strategies, that are self-verifying. Again, in Newman's view,

> [t]here is, then, a circular quality to the process of doing Jewish ethics. We adopt a concept of the tradition, which grounds our model of ethics and so guides our interpretation of classical texts. These interpretations authenticate our moral choices, which, in turn, reinforce our concept of tradition.[54]

Whether conscious of it or not, each bioethicist chooses a particular stance vis-à-vis the textual tradition, and this choice frames as much as constrains how one reads that tradition and what sort of freedoms and authorities one grants the modern audience, care provider, and patient alike. As will be shown throughout this story, modern bioethicists approach the textual tradition with dramatically different hermeneutics that produce divergent—even contradictory—understandings of what the textual tradition says.

A Story of a Story

The following traces the story of a particular—and peculiar—story: Chananya ben Teradyon's fiery dying and ultimate death. Teasing apart dying from death is purposeful because the story itself separates them. Indeed, the story relates his lengthy demise, overflowing with moments, characters, conversations—and as will be shown, changes of mind—all before he ultimately dies. So many bioethicists, however, see in this story only his death; some see only his resistance to it. Few pause to reflect upon the totality of his dying moments, much less upon the tensions and shifts clearly evident therein. Such narrow readings, such hermeneutics, damage the story and, I suggest, take advantage of the vulnerability of people (physicians, care providers, patients, families) bioethicists ostensibly hope to influence. Indeed, it could be that such narrow readings could promote lethal harm—to people, to the text, and perhaps even to bioethics itself. All this needs to be addressed urgently and carefully.

The story of stories and Jewish bioethics told here follows a peculiar narrative path. It begins thin, dives deep, and returns to the surface. This bell-curve of thin and thick analysis details the story's center where complexity is most evident and problematic. Like this first chapter, Chapter 2 surveys how things got to be as they are today regarding the maturation of Jewish bioethical argumentation and the role of narratives therein. Chapters 3, 4, and 5, by contrast, dive deep and develop thicker analyses. There the reader will come across more references and cross-references, quotes, and citations. This slower kind of scholarship enables reading the sources closely, carefully attending to nuances, obfuscations, and absences—aspects that trouble contemporary Jewish bioethical discourse. The final chapter emerges from the depths to reflect on this project's overall story about the complex interrelationship between narratives and (Jewish) bioethics in general. It is my hope that this story—one of many that can be told—heightens everyone's sensitivities to the problems and power of mixing narratives and norms.

Chapter 2

Narratives, Norms, and Deadly Complications

The Torah—how is it written? Upon white fire, in black fire.[1]

Introduction

Judaism begins with a story. It is a fantastic story about the genesis of the world. It overflows with awesome deistic powers, curious creatures of all stripes, conflicts between characters, and endless intrigue. This story has been told and retold for thousands of years and now enjoys global recognition because other religious traditions now claim this story to be their own.

Though commandments appear early on, it is only later—much later—in that original story that any semblance of law comes into the picture.[2] Law's late arrival is telling. It bespeaks the fact that law is intelligible only within a larger narrative framework. And it suggests that despite our desire to focus if not fixate on law as guide and goad for our individual and collective behavior, we should nevertheless give due attention to narrative since it is a larger story that we individually and collectively emerge into, contribute to, and exit from. Our normative life makes sense only within a narrative context. We ignore stories at our peril.

But if we should pay attention to stories, how should we frame our attention? Are there different ways we can appreciate narratives, and if so, which are best suited to help us thrive? By what standard can and should we evaluate these ways? And are there special ways that stories can and should be read in regard to bioethics and to Jewish bioethics in particular? This book offers some suggestions in regard to this last question and may, *en route,* point toward potential responses to the former ones.

So what is a narrative or story? At its most basic, a narrative represents select events within a larger interpretive context. Whereas chronicles and annals merely list events as discrete items, narratives connect them and thereby render them intelligible. Narratives thus juxtapose and explain events in light of each other. Being comprised of two or more events also means that narratives represent a temporal dimension: some time must pass therein.[3] Mary Jean Walker concisely defines narratives thus: "Narrative form involves selectivity and interpretation, and is guided by the needs of meaning and intelligibility such that the meaning of each element arises from its context within the whole."[4] Because of this, narratives are inherently ambiguous, "open-ended and subject to multiple interpretations,"[5] hence their innate attraction.

Certainly narratives come in many sizes and forms. In bioethics they have usually appeared as cases, and casuistry has been credited for reinvigorating the field.[6] But not all narratives employed within bioethical discourse are cases per se. Indeed, some are not even based on real people or events but are concoctions of someone's imagination. Therefore some are abstract, others brief, and many defy easy categorization. So how might bioethicists use such a wide array of narratives?

In her introduction to the now-classic *Stories and Their Limits: Narrative Approaches to Bioethics,* Hilde Lindemann Nelson outlines five ways bioethicists can engage stories. The first is to *read* them, and through one's immersion in the sweep of a story one comes to imbibe whatever morals may be therein. Second, one can *tell* a story, which perforce requires the author to select salient details that promote the moral or worldview one desires. Third, one can *compare* stories as one would with cases, and through such juxtapositions identify overarching paradigms and principles as well as analogies and exceptions. Fourth, one can *analyze* stories from within so as to discern how narratives pit and depict power asymmetries, and frame and navigate moral conundrums. And finally, one can *invoke* stories, pointing to them to reinforce a point, offer caution, or illustrate via parable.[7]

Each of these reading strategies pays attention to details. Whereas champions of other kinds of ethical reasoning, say Kantian deontologicalism or Millian utilitarianism or Beauchamp and Childress's principlism or their like, generally consider details distracting and eschew them for muddying moral calculations, this very allergy renders such modes of reasoning vulnerable to the criticism that they are aloof, too abstract. They "float," in Nelson's terms, above reality; so unworldly, they become unfit for addressing the peculiar thorny details meddling in actual real-life situations. However heuristically interesting they may be, such philosophical modes of ethical reasoning fail in the face of the brambles of lived life where details reside, poke, and prod, and where the messiness of moral decision-making actually occurs.

Not all details are the same, to be sure. It would be as misguiding to fixate on the idiosyncratic at the expense of the context or the overall flow as it would be to note only broad strokes and ignore nuances. Reading too narrowly may overly emphasize certain details whose relevance to the moral conundrum at issue may be nil. Thus, reading itself becomes an ethical exercise insofar as it entails carefully attending to details, evaluating them against each other to find their relative moral import, and assessing those relevant pieces to the salient features of a contemporary situation. For example, while it may be interesting to note that a balloon nearby a car accident spot is bluish-green like the Sea of Marmara, fixating on that detail distracts attention from the larger, more morally pressing question of what has happened within that accident. Indeed, some readers may be so enamored by that balloon's color quality that it colors, literally and figuratively, their reading (i.e., interpretation) of what happened in the accident. That otherwise superfluous detail shades all else.

Conversely, reading too broadly also entails problems. Investigators—readers—of vehicular accidents could gloss over the possibility or fact that a balloon had bounced into traffic and caused havoc, figuring that since it was not in the immediate scene its significance was negligible. Details matter. Discerning which details are significant—and why—is our constant challenge as readers of life and of texts.

For this reason it is important for readers of stories to justify their reading strategies. Readers of stories need to make transparent how and why they do or do not attend to certain details of a narrative and not others, or why they disparage narratives altogether. This is part and parcel of the process of clarifying one's hermeneutic or reading strategy.

Acknowledging that one indeed employs a hermeneutic—and we all do, whether we are conscious of it or not—when consuming and deploying stories in persuasive arguments such as bioethics, necessarily distinguishes three nodes or perspectives involved in ethical discourse. There is the author who reads a text, the narrative of the text the author is reading, and the audience to whom the author addresses her argument and reading of that specific text. These three—author, text, audience—may never meet in person but they nonetheless interact in ethical and bioethical discourse. Narrative theorist James Phelan thus calls for greater care to be given to the "recursive relationships among authorial agency, textual phenomena, and reader response, to the way in which our attention to each of these elements both influences and can be influenced by the other two."[8] For how one reads a text is as much a factor to the bioethical argument as how one thinks of oneself and the nature of the audience one ultimately desires to influence, not to mention what that text is in itself. In regard to Jewish bioethics, we therefore need to question who bioethicists are and whence their authority; we should wonder what the text they cite actually says; and we should inquire how readers might respond to what authors say that text says. What happens, for example, if a bioethicist claims a text says something it does not? Where does that leave the reader? What does it say about the bioethicist? What does it do to the text? Who should protect a text's integrity? What kind of responsibility does an author have to an audience that does not have access to the original text? How should an audience respect and respond to an otherwise revered bioethicist who says the text says something it does not?

There are further questions complicating the interrelationships between author, audience, and text—especially when we take narrative texts seriously. In addition to justifying how they read a particular story, authors also need to defend why they look to that particular story at all. The story's relevance to the moral conundrum at hand may or may not be readily apparent. What about other stories—how does an author use them? Are they mentioned, alluded to, studied in depth, glossed over, or completely ignored? Why? Such questions pertain to the issue of canon, for when bioethicists turn to narratives A and B and not C they perforce construct the organic boundaries of which stories are to be considered relevant for certain bioethical conversations and which should not be so involved. But consumers of a bioethical discourse need to be wary of such canons; perhaps narrative B has little to say about a particular topic despite what bioethicists insist otherwise, and story C really should be considered but hitherto has been ignored. This realization provokes meta-narrative questions that need addressing: if multiple stories are relevant to a particular issue, how are the decisions made that one of them or a select few are more salient than the others? By what criteria do authors privilege certain stories over others? And from where do those criteria come? And when the selected stories point to dramatically different conclusions—who decides which story should trump and thus provide normative guidance, and on what grounds is that decision made? Louis Newman summarizes these concerns well: "When different narratives point in different directions, we are left without even the (relatively ambiguous) moral guidance that a single story can provide."[9]

Newman points to another issue about reading stories. Even if we give bioethicists the benefit of the doubt and agree that the stories they invoke are indeed relevant to the present issue, we still need to examine how they read them and why they read them in the ways they do. That is, how do bioethicists handle the multiplicity inherent within narratives? A bioethicist cannot but become entangled in narratives' open-endedness.[10] Stories are by nature ambiguous, and bioethicists—themselves readers of those stories—must wrestle with that ambiguity. As Peter Knobel says, "there are aspects of stories that can guide us into different *ways* of looking at a problem, and not just for a different result."[11] Some authors embrace ambiguity and bring it forward into their normative bioethical arguments. Most, however, strip stories of their potential polysemy, cutting and reframing them so that only one interpretation or meaning emerges. They eliminate the "ways of looking" and say a story proffers but one way of conceiving a problem or its solution. This impulse to restructure stories so that they countenance singular conclusions paradoxically points to the fluidity narratives bring to normative discourse and the vague boundaries between what they might endorse and what they might proscribe.[12] In brief, just as narratives themselves are composed of selected and interpreted events, readings of those narratives are also nothing but selections and interpretations; readings *are* narratives.

Such questions and caveats both complicate and make interesting the story of how narratives fit in modern Jewish bioethical discourse. Their role can be understood in light of a larger set of questions pertaining to the relationship between and role of narrative in Jewish law. This broader basket of concerns has long concerned Jews; indeed, it is as old as the rabbis. It is not our intention to rehearse the totality of this lengthy debate or to weigh in on it; rather, in the next section we restrain ourselves to highlight only some of the more salient features therein. We then turn our sights to modern Jewish bioethics and its troubled relationship with narratives. As will be shown, stories have long been incorporated in bioethical deliberation though they have not been treated very well. Indeed, most bioethicists appear to have the predilection to point to stories to support their positions on a specific moral issue. Few spend time or effort to plumb the stories for their descriptive richness or prescriptive diversity. Though this reading strategy is lamentable, it becomes potentially lethal when we consider modern Jewish bioethical deliberations regarding care at the end of life, specifically regarding euthanasia. For better and for worse, bioethicists writing on euthanasia read—or misread, as it will be shown—stories, especially a specific one, and in so doing they endanger not only patients already vulnerable to power asymmetries and ill health, but also families of those patients, their physicians, fellow clergy who may or may not be willing or able to disagree with the bioethicists, and others. Understanding how bioethicists read this story and make their normative recommendations is thus a lethally relevant task the urgency of which is all the more pressing with the growing number of people involved with their own or others' dying processes.

NARRATIVE AND LAW

Long has the assumption been held that what is known as halakhah and aggadah are two distinct genres. Their connection was all but nil, their relationship nonexistent. Drawing inspiration perhaps from the 1923 essay "Law and Legend" by Hayyim N. Bialkik, the great scholar of Jewish legends and lore, scholars in recent decades have begun to question this assumption. Indeed, the increasing acknowledgment that aggadah and halakhah intertwine in classic texts has inspired a burgeoning conversation on the truly complex relationship between these genres, and questions arise whether they rightfully should be teased apart even for heuristic purposes.[13]

The impulse to disconnect story from law, aggadah from halakhah, emerged early on in rabbinic Judaism.[14] For example, R. Zera (a third–fourth century Palestinian amora) taught in the name of Samuel that rules should not be based upon anything except for the Talmud, and certainly not on aggadot (the plural of aggadah).[15] Not only was aggadah defined negatively—as everything that wasn't law[16]—but it also was further demoted from having normative suasion when the early medieval Geonim in the tenth century opined that "aggadah should not be relied upon" because stories are generally inconclusive and are conjectures.[17] This spurred later scholars to study these texts separately and for different goals. Law was plumbed for normative precedent and jurisprudential reasoning. Stories were appreciated for revealing sociopolitical insights of various historical contexts. While the former were lauded for shaping Jewish behavior, the latter were all but dismissed because, for all intents and purposes, they only reflected Jewish character. Serious scholars studied law; others bemused themselves with stories.

Still, aggadah is not midrash. Midrash, as commonly understood, is a mode of interpreting biblical material. For millennia, Jews have interpreted The Bible and often compiled the interpretations into distinct collections. Some of these are midrashic interpretations of words, some of stories, and some of laws. Even though midrash is closely linked to The Bible, this does not preclude a relationship between aggadah and biblical themes. Indeed, scholarship suggests that aggadah's task is to elucidate God's ways and mysteries, and more.[18] Since aggadot do not share a single identifiable characteristic or style, they should be considered less for their form or genre than for their content. Aggadah—rabbinic stories—illumine aspects of theology, realia, and character that other genres like law and midrash cannot. If one needs a single word to encompass aggadah, I suggest lore, as lore means that which is taught, a doctrine or teaching, advice and counsel, and even a body of knowledge—indeed, lore is something that can be learned.[19]

That said, the apparent difference between halakhah and aggadah—between law and lore—should not cause real schism today.[20] First, law and lore are inextricably intertwined in the biblical and Talmudic corpus, making it all but impossible to tease them apart—despite medieval efforts to do so. Second, in regard to the Talmud at least both kinds are authored and populated by the same people. The sages not only rendered legal decisions, but they also spoke of themselves doing so—among other things. That is, aggadah and halakhah are *by* and *about* the same people. Relatedly and third, insofar as aggadah bespeaks the peculiarities of existence and the dimensions of *au courant* culture, they put flesh on legal bones. They animate what otherwise would only be a skeletal depiction of classic Jewish life and theology.[21] Lore and law may not be commensurate but they certainly are complementary; they implicate each other, or, as Bialik says "they are merely two phases of the same phenomenon."[22] They enhance each other.[23]

This is not a new argument, to be sure. Robert Cover famously made the argument that although law and narrative are distinct, they nonetheless are interrelated; law cannot be fully comprehended without the stories that render them intelligible in the first place.[24] And philosopher John Arras concurs: "For no matter how far we progress toward the ethereal realms of principle and theory, we ought never to lose sight of the fact that all of our abstract norms are in fact distillations (and, yes, refinements) of our most fundamental intuitive responses to stories about human behavior."[25] Indeed, it could be reasonably argued that the dark concreteness of laws, norms, and principles are indecipherable without appreciating them against the lighter backdrop of larger narratives.

Or as the Jewish textual tradition might put it, black fire is intelligible only when juxtaposed to or superimposed upon white fire. The Torah and Talmud, for example, are black ink upon pale parchment. Though Cover promotes the notion that a (biblical) master narrative undergirds all (even secular) law, this may not be the case for halakhah and aggadah since aggadot are more piecemeal, fractions of stories, and not of a whole.[26] Even if there is no overarching, shimmering grand narrative behind aggadot, studying law and lore together seems to be the best way forward.

There are additional reasons for this plan. Sociologically speaking, contemporary Jews are increasingly disinclined to "feel compelled by a halakhic ethic."[27] And the technical jargon and picayune argumentative strategies so prevalent in halakhic discourse make it a difficult genre to fully appreciate by readers untrained in the ins and outs of rabbinic legal reasoning.[28] Inasmuch as bioethicists seek to persuade audiences to do according to their assessments of a particular moral conundrum, it behooves bioethicists to argue with readily accessible rhetoric. Legal reasoning no longer suffices on its own; stories must be considered and deployed for Jewish ethics to remain relevant and compelling.[29]

NARRATIVE AND BIOETHICS

This holds for Jewish bioethics, too. The turn to aggadah in bioethical discourse is a relatively recent one in the history of the field. Although the field is not old, its development nevertheless evidences maturation, intrigues, and deepening fissures vis-à-vis aggadah. The role of the story within Jewish bioethics is itself a story.

In his magnum opus *Biblisch-talmudische Medizin* (1911), Preuss includes among the primary sources from which he draws his medical archeology several collections of midrashim: *Mekhilta* (legal midrash on Exodus), *Sifra* (legal midrash on Leviticus), *Sifre* (legal midrash on Numbers and Deuteronomy), *Midrash Rabbah* (exegetical midrash on the Bible), *Midrash Tanhuma* (homiletic midrash on the Bible), *Pesikta de Rav Kahana* (exegetical midrash on the Bible), *Midrash Tehillim* (exegetical midrash on Psalms), and *Yalkut Shimoni* (midrash on the Bible).[30] And he certainly cites and recites aggadot from other rabbinic materials, especially from the Talmud itself. Preuss astutely observes that these sources—as well as the legal ones—are not medical tracts; indeed, medicine and medically relevant content appears but infrequently within them. Nor are any of these texts written exclusively by physicians. Thus the best that can be said of these sources is that "the majority of [their normative] pronouncements belong to folk medicine."[31]

Folk medicine, Preuss notes, differs from scientific medicine in that it "imparts suggestions, methods and data" to the latter, and scientific medicine is based on rational observation and the evaluation of facts.[32] Yet clearly distinguishing between these kinds of medicine in Jewish sources—the Talmud in particular—is problematic.

> [As] soon as we begin to separate science from folklore, we search in vain for criteria whereby this separation can be accomplished. Most Talmudic teachings came down to us anonymously through tradition, and where the name of the author is mentioned, we often learn nothing of his profession and his life. Or, coincidentally, the name of a physician may be mentioned, but nothing is stated about his medical teachings. For the Torah and the Talmud, as must be reiterated, are primarily legal sources, and not medical textbooks.[33]

Moreover, since the Talmud itself comprises many centuries of rabbinic deliberation, it should be assumed that medical knowledge advanced even within its pages. Given the difficulties in pinpointing with any precision who said what, when, and where, it is all but impossible to ascertain

> the appropriate conditions which were contemporary with the teaching, within whose framework a real picture can be elucidated. For undoubtedly—we do not have to go into detail—medical science is influenced by the actual conditions of the culture, just as it exerts its own influence on the fashioning of many outward life situations.[34]

The historian of medicine is thus obliged to avoid reading these classic sources as would the religiously devout, lest one apologize for what cannot withstand reason's scrutiny. "*For historical research,*" Preuss emphasizes, "*religious sentiments should play no role at all;* only the facts must speak for themselves."[35] For all too often one might be tempted to fill in gaps or questionable material with "the product of one's own imagination" and present it "as historical fact."[36] Such fantastical eisegesis pollutes good scholarship; it can corrupt bioethical deliberation and condone problematic, injurious, and even lethal, medicine.

Preuss's *Medicine* thus cites aggadah without retelling the stories in any great detail. He therefore models a kind of reading that points to narratives as evidence of biomedical knowledge, and he does not plumb the stories themselves for nuance, depth, or insight about the morally troubling aspects of medicine. For example, in the section on suicide, he records "a rather rich casuistry on suicides" in the Gemara, at the end of which he places the story of Chananya ben Teradyon—to illustrate again Judaism's general repulsion of self-destructive behavior.[37] According to what he says, the story augments his argument without furthering it. His two-sentence snippet of this story paradoxically undermines his preferred historiographical methodology, for in so doing he imposes upon the story what he sees as medically relevant therein. (We examine the precise wording of his invocation of this story in subsequent chapters.) Though it is beyond the scope of this book to adumbrate the many stories he tightens to fit his project, let this suffice to demonstrate his impulse to point to and not plumb narratives.

To be fair, Preuss did not imagine his work to be bioethics per se. His was an exercise in medical archeology, indeed anthropology—since he endeavored to show interrelations between ancient, rabbinic, and medieval Jewish medical knowledge and the medical knowledge of surrounding cultures and traditions.

The next great contribution to the field was Immanuel Jakobovits's *Jewish Medical Ethics* half a century later. Perhaps inspired by Preuss's historiographical leanings, Jakobovits organized his work so that each topic includes a traipse through historical—and comparative—layers of Jewish medically relevant literature. Elie Munk, in his preface to Jakobovits's volume, opines that by historically situating the text, Jakobovits "permit[s] the solutions inspired by Jewish ethics to appear in their true light."[38] That is, the best way to appreciate ancient sources is to contextualize them in the peculiarities of their historical moments—and this is no less true for medicine. Munk then goes on—listing with eerie prescience and wisdom a range of issues that provoke moral outcry to this day—to insist that that historically true light of Jakobovits' oeuvre is medically and morally truly right:

> Thus, the vast subjects of medico-moral conflict (such as euthanasia, dissection, abortion, the problems of eugenics and legal medicine, as well as many others), which have not ceased, over the generations, to beset the moral conscience of humanity, are treated in the light of the eternal truths whose authentic source is the biblical revelation. The positions taken [by Jakobovits] are clear; they are not the result of personal, logical or sentimental

> considerations, always subject to caution, nor of systematic philosophy, a product of the human spirit and, consequently, always relative in value; they are based, in the final analysis, on the solid foundations of universal morality whose charter is contained in the Decalogue. In practical life, man, faced with a crisis of conscience, can give it his confidence.[39]

Baruch Brody would take issue with Munk's desire for all humankind to hearken to Jakobovits's rendering of Jewish medical ethics. For in Brody's view, claiming that Judaism "mandates a certain solution to a problem in medical ethics, where the claim is that this solution ought to be adopted, according to the Jewish viewpoint, by society in general"—falls in the face of the fact that the vast majority of Jewish (normative) texts speak of duties obligatory upon Jews and not upon gentiles.[40] Even if one construed the Noahide Laws expansively, it is still difficult to make that claim much less enforce Jewish norms upon non-Jews.[41] Of course, we could be misreading Munk, for perhaps he meant "*Jewish* man, faced with a crisis of conscience" can rely upon Jakobovits for unblemished bioethical guidance—that is, the book is meant to clarify Jewish norms for Jews alone. But given Munk's insistence that the Jakobovits' scholarship is based on "universal morality," it seems he would rather it serve as a moral and medical desiderata for all, Jews and gentiles alike.

Munk's confidence in Jakobovits's work is both evident and infectious. Scores of Jewish bioethicists—and rabbis and laypeople—turned to his *Jewish Medical Ethics* not as the preeminent piece of scholarship in a young field, but as the definitive and conclusive word on particular topics. Such enthusiasm, however, glosses over Jakobovits's own observation that the primary audience of earlier Jewish medical tracts, especially of *responsa,* was not the physician but the patient who posed, or is, the question in the first place.[42] Implied behind this observation and inadequately acknowledged by Jakobovits is that what drove Jewish medical thinking up to this point in history was not the deductive application of principles to cases—as was the wont in Catholic and Islamic medical ethics, but the inductive reasoning from the peculiarities of idiosyncratic cases to larger principles and rules, and to older similar yet different cases. Stories and personal narratives of actual people fueled the field—even though few if any acknowledged this.

Jakobovits nevertheless takes pains to appreciate the historical nature of the Jewish sources he references. He notes that the medieval codes, while impressive in their breadth and depth, do not reflect *au courant* medical knowledge as much as they do the knowledge ensconced in the Talmudic texts they condense: "In their contents and largely in the illustration of their legal principles, therefore, they mirror the conditions of Jewish life and law at the time of the Talmud rather than those prevailing in their own days. This consideration applies in particular to the medical data contained in these codes."[43] Retrospection also holds for the *Shulchan Aruch,* the most revered of the medieval law codes, composed by Joseph Karo in 1563, for in it "little or no account is taken of medical facts observed in the intervening millennium [between the Talmud and it], relatively insignificant as these may have been."[44] Any alteration of medical practice had to be justified through some other means than citing earlier sources. Rather, innovation had to be justified outside the textual tradition, not from within it: "The rabbis, to justify such modification or invalidation of the original laws, argued that they had lost their validity owing to changes in time, local conditions or even nature since their enactment."[45] Or, in a worst-case scenario, it could be argued that the earlier sages were deficient or erroneous in their knowledge.[46] In short, if a rabbi writing on medicine wanted to promote a procedure other than what was recorded in the literature, he had to concoct some kind of story to justify the alteration.

Viewing earlier medical texts with suspicion is thus reasonable—for they may not reflect the sociopolitical circumstances or technological advancements of their provenance, and whatever novelties they promote may themselves be reliant upon questionable rationales. This kind of skepticism has early medieval roots, as Jakobovits rightly notes. The tenth century sage Sherira Hai Gaon warned that since the earlier sages were not physicians, their prescriptions regarding medicine should not be considered law, nor should contemporary law rely upon them. "You must not, therefore, rely on medicines mentioned in the Talmud. Only he may use them who has had them examined and confirmed by experienced physicians, and who has the assurance that at least they can do no harm. Thus our forefathers also teach us that one may employ only those remedies of which one is certain they produce no injurious effects."[47] Such caveats lead Jakobovits to conclude that early sources "do not necessarily reflect the outlook of the time of their composition, and that the medico-religious views recorded in them are, in certain respects, flexible enough to allow for continuous revision and adjustment."[48] Whereas Jakobovits sees in prior texts wiggle room for interpretation and ample room for advancements in scientific data, Munk sees in Jakobovits the definitive word, decisiveness, and conclusion. For one the conversation continues, for the other nothing more can or should be said.

Still, how does Jakobovits treat stories as they might pertain to medicine? Again, since we cannot survey here the totality of his *oeuvre,* take the story of Chananya ben Teradyon dying by fire as an example. Jakobovits cites this troubling narrative in a section entitled, "Treatment of the Dying." There, citing the same pieces of the story as does Preuss, Jakobovits rules that the "uncompromising opposition to any deliberate acceleration of the final release is well exemplified by [this] sage."[49] (More will be said of his precise quotation of this story in subsequent chapters.) The story does not serve as the source for the norm against hastening another's death; it merely exemplifies or illustrates that norm. It supports his argument, or so he says.

Note the shift. It is fascinating that for Preuss this story bespeaks of suicide and the allergy Judaism has toward it, yet for Jakobovits it pertains to hastening someone else's death, not one's own. This dramatic difference of interpretation is all the more astonishing because they point to *exactly the same* two sentences of the (much larger) story. Point, and do not plumb. They appreciate the story neither in its full expression (the whole *sugya*) nor in its larger Talmudic context (the chapter in which it is found). The story—or to be precise, this tiny bit of the story they cite—indicates classic support for the divergent positions they champion. Or so they say.

Indeed, they say a great deal when they say so little about and with this story. They say that this story can be condensed into these two sentences. This perforce means these two sentences are the story's essence. It suggests nothing else in the story should interest bioethicists and certainly nothing else has normative import. Essentializing the story—or any story, for that matter—is an intellectually fraught endeavor. It remains unclear why these scholars choose these two sentences as the story's essence. By what criteria did they make this selection? What disqualified or demoted the rest of the story—and be assured there is a great deal to the rest of the story—to the margins, indeed, to the dustbin of bioethical inquiry? Such questions approach the problem from the author's perspective. More questions arise from the perspective of those who consume their bioethical arguments. For those of us unfamiliar with this Talmudic story, why should we be deprived of the larger narrative? Why should we be led to believe that these sentences are truly bioethically important? What should we make of the disconnect between one author's proscription of suicide and another's against euthanasia—both positions hinged as they are on this singular narrative? Should we ignore the totality of the story only because these rightfully lauded scholars present just this snippet? How is such deference a health-promoting attitude, or does it carry some danger especially since the topics to which they apply this story pertain to life and death? This book attempts to address these and related questions.

Still, for better and for worse, Preuss and Jakobovits model a reading strategy that then prevails for several more decades among Jewish bioethicists. This kind of selective reading or essentializing became the norm in the field. This claim finds support in Louis Newman's overview of the field of bioethics.[50] He identifies three major models therein: (1) the "legal model" in which legal reasoning follows formal rules and the textual tradition is relied upon to supply normative precedents; (2) the "covenantal model" wherein the dialectical nature of the covenantal relationship between God and humans offers normative guidance; and (3) the "narrative model" that extracts guidance from larger, master narratives and the open-ended and imaginative interpretations derived therefrom. Though the last is the least developed kind of modern Jewish bioethical discourse, on the whole all three styles of bioethical argumentation read stories as do Preuss and Jakobovits. All point to snippets of stories supposedly supporting the positions they endorse.[51] As these three models of bioethical argumentation have come to dominate the field, a few moments should be devoted to clarifying their features.

Formalist

The formalist school, mostly populated by Orthodox scholars, in which laws theoretically rule supreme, nonetheless invokes stories when constructing normative arguments. Two prominent contributors will suffice to illustrate this trend. Dr. Fred Rosner (1935–) is perhaps the most prominent scholar in the field of Jewish medical ethics, having published nearly 800 articles and, we should note, translated Preuss's magnum opus into English. Throughout his *oeuvre* he carefully reviews classic and contemporary sources pertaining to his subject matter. Though his wont is to summarize others' arguments and positions, he also recapitulates in part or in full critical classic sources when they are pivotal—including narratives.

For example, in his summary of Judaic perspectives regarding contraception, he identifies four Talmudic techniques used by women to prevent conception: the rhythm method, violent movement postcoitus, oral contraception, and a spermicidal absorbent material.[52] In regard to oral contraception, he cites the story of Judith, the wife of R. Chiyya:

> Judah and Chizkiah were twins. The features of one were fully formed by the end of nine months, and those of the other were fully formed by the beginning of the seventh month.[53] Judith, the wife of R. Chiyya, had suffered greatly during childbirth, changed her clothes and appeared before R. Chiyya. She said to him, "Is a woman obligated to procreate?" He said to her, "No." She returned and drank a sterility potion. When this was revealed to him he said to her, "Would that you had bore me one more issue!" A Master taught, "Judah and Chizkiah were twin brothers, Pazi and Tavi were twin sisters."[54]

According to Rashi's commentary, both sets of twins—Judah and Chizkiah, Pazi and Tavi—were children of Judith and Chiyya.[55]

Certainly the story endorses the notion that women are not duty-bound to fulfill the obligation to procreate (and however much others might argue otherwise, procreation is not a commandment—mitzvah—but a blessing, according to Genesis 1:28). This thus opens up the possibility and permissibility for women to use a contraceptive; here, it is an oral one. Yet on the other hand, when read in its totality, the story bespeaks an anxiety about women's sterility and women's control of their reproduction generally. More specifically, it reflects Chiyya's anxiety. Though Chiyya thinks women are more bitter than death (following Ecclesiastes 7:26), he nonetheless takes pains to shower his wife with gifts whenever possible because, as he says to his colleague Rav, "It is sufficient for us that they raise our children and save us from sin."[56] Obviously, his anxiety should be taken with a bit of skepticism since in all likelihood the Talmudic text was composed by men and geared for men. It thus could be understood as a story that simultaneously permits oral contraception for women and bemoans this very permission. Not everything permissible is always widely desirable.

The complexity of this story is lost, however, in Rosner's recitation of it. He starts his quote of the story with "Judith, the wife of R. Chiyya . . . " and ends it after she drinks the sterilizing potion. In so doing Rosner silences the fact that Chiyya and Judith already had twin boys, and that they would also have twin girls, and he also suppresses the seemingly very real emotional turmoil Chiyya experiences. When Rosner concludes his essay with the position that modern oral contraception "seems to be the least objectionable method of birth control in Jewish law," he achieves his goal of articulating a clear normative conclusion—which is sine qua non for the formalist school.[57] Yet his very argumentation strips the classical story of its emotional dimensions and collapses it into what he perceives as its singular normative stand. He thereby removes the humanity of the humans involved—both in the text and, by extension, those moderns he attempts to sway.

Rosner is not alone in this school when he points to stories and selections thereof to support his normative conclusions. Another leading Orthodox Talmudist and ethicist who does something similar is J. David Bleich (1936–). Bleich, like Rosner, squares himself with prior scholars' positions. He explicitly states that the "only" proper way of identifying normative positions is by examining questions "through the prism of Halakhah for it is in the corpus of Jewish law as elucidated and transmitted from generation to generation that God has made His will known to man."[58] Regardless of the issue at hand, ambiguity is shunned in favor of clarity and conclusion—and the legal textual tradition is the sole source. Answers are to be discovered in the law and not created, and they certainly should not be derived from aggadah.[59]

Bleich's commitment to the primacy of law is evident in his scholarship: he recites narratives almost exclusively in his footnotes. Marginalized, as if buried beyond the walls of his otherwise tight legal reasoning, stories nevertheless exist and exert some influence upon his decision-making. Sometimes, however, narratives find their way into his main texts, even while he labors to stay focused on norms and laws. For example, regarding the improbable yet occasional occurrence of conjoined twins, Bleich renders a ruling about the permissibility of surgically splitting them, which would inexorably bring about the weaker child's death, based upon his reading of several stories.

One is the story of Sheva ben Bichri, who apparently merited the death penalty for assaulting King David (II Samuel 20:4–22). The Talmud[60] relates a story of heathens surrounding a group of traveling Israelites and threatens, "Give us one of your company and we shall kill him; if not we will kill all of you." The rabbis debate this and eventually rule that the group should elect to be killed than transfer any one for certain death. But the story continues: if the heathens specify someone, such as Sheva ben Bichri, the group is to deliver him over so the group would survive. R. Shimon ben Lakish favors this conclusion only if the identified individual is already deserving of the death penalty—as was Sheva ben Bichri. R. Yohanan holds otherwise: the person's guilt is irrelevant; that he was singled out by name is all that matters. Both sages nonetheless agree: the designated individual is to be handed over for certain death so as to protect the rest of the group from potential harm.

This story leads Bleich to question the analogy that the weaker twin is akin to the "designated" Sheva ben Bichri. For though the smaller twin may not survive under any circumstances and is thus "designated" for death, "the 'designated' twin is clearly innocent of any wrongdoing and cannot be regarded as culpable for the penalty of death."[61] Thus he warns against making any ruling on this reading alone.

Yet, when he returns to the theme of conjoinment a few years later, Bleich notes that the Talmud's story continues. It speaks of R. Joshua ben Levi who instructs a town harboring a fugitive to turn that individual over to the marauding heathens to protect the town from being fully besieged. Whereas beforehand the prophet Elijah used to frequent R. Joshua's house, he now refrained from visiting the rabbi. R. Joshua fasted and prayed so vehemently until finally Elijah appeared with the retort, "Shall I reveal myself to a betrayer?!" R. Joshua defends himself by saying he was merely following the law, to which Elijah rhetorically responds, "Is this the law of the righteous?" Having reached the *sugya*'s end, Bleich concludes thus:

> Put simply, the Talmudic narrative teaches that delivery of a victim to death, even when halakhically defensible and motivated by an imperative to save others, is not consistent with piety. Although legally defensible, it is a course of action that a person possessing a keenly honed moral sensitivity should eschew as repugnant.[62]

Though he has not altered his ultimate position vis-à-vis the permissibility of cutting apart conjoined twins, he reaches this conclusion based on two different reading strategies *of the same story*—even though he is adamant that stories cannot (ever) serve as grounds for halakhah.

The other story Bleich invokes regards King Solomon and Ashmedai, the prince of demons.[63] Ashmedai brought forth from beneath the earth a two-headed man and stood him before King Solomon. He married and sired several children, one of which also had two heads. When the time came to divvy up his inheritance, the two-headed child insisted on a double portion. The case came before King Solomon, who devised a test to discern whether the two-headed child was a single person or two. Solomon had one head covered with a cloth and poured hot water upon the other. If the blinded head cried out, he knew the conjoined twins were a singleton; if no response emerged, they were two. As it happened, both heads cried out in pain, and so Solomon ruled that the conjoined twin could rightfully receive one inheritance portion. Bleich insists that the story is inconclusive halakhically even though narratively it points to a particular perspective.[64] But then he turns to the sixteenth-century commentator Bezalel ben Abraham Ashkenazi, who extracted this norm from the story: conjoined twins "are a single offspring with two heads, and they shall be judged only as a single person."[65] Yet in Bleich's view this means "conjoined twins, each endowed with a full complement of organs, are clearly separate individuals, regardless of whether or not they respond autonomously to pain stimuli."[66] Despite the fact that Bleich obviously contradicts this earlier halakhic scholar, what is more fascinating is that he turns to and leans so heavily upon such stories—and fantastical ones at that—when rendering his bioethical normative positions, and given his own predilection to adhere (nominally, at least) to "the principle that halakhic matters are not subject to determination on the basis of aggadic statements."[67]

Another story Bleich invokes is the story of Chananya ben Teradyon, but only partially so, in a lengthy note on his examination of Moses Isserles's ruling about moving an otherwise moribund patient.[68] Much more will be said about this particular story and Bleich's reading of it in subsequent chapters.

Covenantal

Covenantal ethicists, no less than their legal formalist colleagues, are also prone to incorporate stories when constructing norms. This school grounds ethics in the notion of covenant, the goal of which "is to live faithfully in relationship to God and in continuity with earlier generations of Jews who likewise have attempted to live faithfully in this same relationship."[69] Without either replicating what previous generations did or transcribing ancient rules into modern milieus as might legal formalists, covenantalists endeavor to discern and creatively balance their competing responsibilities within their understanding of that foundational relationship with God. Undisputedly the most prominent covenantal theologian and ethicist (though admittedly not a practical ethicist as such[70]), Eugene Borowitz (1924–) has written that both aggadah and halakhah inform covenantal deliberation. Both are necessary to develop a "Covenantal Dialectic," a dialectic that fully appreciates the broad range of possible patterns of Jewish action. In regard to law, Borowitz points to the fact that there are decisors who are stringent (*machmirim*), who would "let the law make its fullest demands upon us," and there are also more lenient ones (*mekilim*), who recognize "that the law was not meant for angels but for human beings with all their limitations."[71] The very existence of these differing attitudes toward law endorses the "dynamism reflecting the Covenantal Dialectic."[72]

Aggadah also contributes to this Covenantal Dialectic inasmuch as it articulates the importance of human experience in the throes of history. For instance, Borowitz points to the various aggadot—many of them midrashim, actually—regarding the revelatory moment at Mt. Sinai.

Again, the rabbis teach [1] that God's love rather than Israel's merit brought God to choose them to receive the Torah and be God's own special people. [2] Or they say that God legally acquired them by redeeming them from Egypt or, [3] when they hesitated to accept the Torah at Sinai, coerced them into doing so by lifting the mountain over their heads. So classic Jewish theology ruminates on the divine prerogatives. But it also does so from the perspective of Israel, perhaps the most famous being [4] the legend of God unsuccessfully peddling the Torah to the nations until Israel joyfully shouted, "We shall do it and hearken to it." [5] So too the Sinai coercion story has a rejoinder from the human side, that in the days of Mordecai and Esther the people willingly accepted what had once been forced on them.[73]

In his famously succinct style, Borowitz indicates without reciting—or even citing—classic rabbinic narratives, weaving them together to create a tapestry to his liking. But first, we should consider what Moses says:

For you are a people consecrated to the Lord your God: of all the peoples on earth the Lord your God chose you to be His treasured people. It is not because you are the most numerous of peoples that the Lord set His heart on you and chose you—indeed, you are the smallest of peoples; but it was because the Lord favored you and kept the oath He made to your fathers that the Lord freed you with a mighty hand and rescued you from the house of bondage, from the power of Pharaoh king of Egypt.

(Deuteronomy 7:6–8)

It is unclear why Borowitz insists that it was the rabbis who taught [1] that God chose Israel for a special relationship; Moses said this. That said, he goes on [2] to recite a third-century parable of Israel and God, in which God is likened to a human king who asks the people if he may reign over them. "Have you given us any benefit that you should reign over us?" they inquire. So the king builds for them infrastructure, secures their sustenance, and wages war on their behalf. He again asks for their permission to reign and they grant it. So it is for God and Israel, the story explicates: God extricated Israel from Egypt, fed them in the wilderness, defended them against Amalek, among other things. Thus when God asks to reign over them, they approve.[74] The third [3] is a Talmudic story exploring the meaning of the biblical phrase, "*they stood underneath the mountain*" (Exodus 19:17). The rabbis interpret this to mean that God held Mt. Sinai above the Israelites like a casket and said, "If you accept the Torah, well and good. But if you do not, there will be your grave."[75] The fourth [4], perhaps medieval, story (itself an agglomeration of other stories) depicts God revealing revelation in all the languages of the world yet every nation rejects what God demands; only Israel survives hearing revelation and takes upon themselves these divine strictures.[76] Borowitz then weaves back [5] to the third story about the overturned mountain, which states, "In the days of Ahashuerus a generation accepted revelation, as it says, '*The Jews established and accepted*' (Esther 9:27), which means they established what they had already accepted."[77] Borowitz needles through these stories to thread his own. He stitches them into his overarching argument that modern Judaism—no less than previous iterations of Judaism—necessarily weaves contradictions, tensions, and alternatives into a lively patchwork in which individuals *as* individuals ultimately determine "what to make of God's demands and Israel's practice, tradition, and aspiration."[78]

The power of narratives to create the terrain for Jewish choices and living is expanded upon by another covenantal theologian, Rachel Adler (1943–). In her ground-breaking feminist tract *Engendering Judaism: An Inclusive Theology and Ethics,* Adler takes a cue from Robert Cover's work on narrative's grounding for law. Insofar as law cannot be established upon coercion, it "must claim legitimating moral qualities, which it locates in its constitutive narratives."[79] These foundational narratives, however, do more than articulate positive moral qualities. Cover notes, as "every legal order must conceive of itself in one way or another as emerging out of that which is itself unlawful,"[80] these foundational stories also serve as precedents for transgressing the very norms they support.[81] Because narratives can both ground and undermine norms they perforce are threatening, and this potency is a key piece of their fascination. *Engendering Judaism* goes on to demonstrate narratives' dual nature of grounding and destabilizing norms, with Adler reading numerous stories as they pertain to taking women seriously in the first place when constructing Jewish norms, invoking and hearing women's voices in liturgical settings, reformulating a Jewish sex ethic in which partners are celebrated regardless of their gender, and—most famously—reconstructing marriage as a partnership between lovers and not as a business exchange of acquisition.

Narrative

The third type of modern Jewish bioethical discourse Newman identifies is what he calls the narrative school. Here stories are plumbed for their content and are appreciated for their form. "The very structure of the story—its characters and plot, its dramatic or tragic or comic qualities—are part and parcel of the message."[82] Iterative encounters with certain stories enable Jews to "internalize their structure, to project themselves into the world described by the story, indeed, to see the world around them as a continuation of the story."[83] What is inscribed in ancient ink thus becomes embodied and enacted in modern lives. The story literally shapes behavior; it molds norms into being.

Perhaps the most avid proponent of Jewish narrative ethics is William Cutter (1937–), and he demonstrates this narrative "way of thinking" most vividly in an essay on euthanasia. There he urges turning to stories instead of law because stories reflect "the real and lived experience of the patient in home or hospital," and they can cue "how we might make decisions regarding the last stages of a loved one's illness," which requires being "nearly as rigorous with our use of stories as we have been with our application of 'halakhic formalism'."[84] The story he examines in this instance relates the dying moments of Judah HaNasi (also known plainly as Rabbi), the second–third-century Palestinian tanna who redacted the Mishnah, who was known for suffering a debilitating digestive illness. Here is the story in detail:[85]

1. On the day Rabbi died, the rabbis decreed a fast and[86] prayed for mercy. They said that anyone who said that Rabbi died would be pierced by a sword.
2. The handmaid of Rabbi ascended to the roof[87] and said, "Those above [immortal angels] request Rabbi, and those below [mortal humans] request Rabbi. May it be [God's] will that those below overturn those above."
3. When she saw how many times he went up to the privy, and took off [his] *tefillin* [prayer phylacteries], and put them on, and how much he was pained,[88] she said, "May it be [God's] will that those above overturn those below."
4. But the rabbis[89] would not cease from their prayers for mercy; she picked up a vessel and threw it from the roof to the earth.
5. They[90] [the rabbis] ceased from [praying for] mercy, and Rabbi died.
6. The rabbis said to Bar Kapara, "Go, investigate." He went, found [Rabbi] had died, tore his cloak and turned the tear behind him.

7. He began and said, "The Erelim[91] [angels] and mortals grasped the Holy Ark. The Erelim overpowered the mortals, and the Holy Ark has been captured."
8. They said to him, "He died?"
9. He said to them, "You said it, but I did not say it."

In Cutter's view, this is a powerful "minimal story" in that it shifts at one point from one predictable state to a conclusion not otherwise anticipated.[92] He notes that it has become a *locus classicus* among medieval and contemporary Jewish bioethicists precisely for this shift, that is, for its apparent support for removing impediments to death despite the overriding principle of not hastening another's demise.

Such a narrow reading is unsatisfactory for Cutter, however. Narratives cannot be constrained to any one particular interpretation, thus, Cutter insists, we must revisit the story not as a minimal one but in pursuit of "other dimensions and themes, to its rhetorical strength and possibilities for surprise."[93] Now attuned to other details beyond the singular major shift, this second read notes the assumption that others should pray on Rabbi's behalf and the mitigating circumstance warranting the change of behavior—his extreme suffering. What about the role of women in healthcare, and prayer, and witnessing suffering? These, too, merit scrutiny and can receive a much needed boost from this story if one looks at it not for its pivotal "norm" but for its otherwise sidelined details. For there, in the margins and crevices of the story, are suggestions on ways to think about caring for those nearing their ends.[94]

However rich this second read may be, Cutter again returns to the story a third time. Here he asks a larger question, one that transcends norm and detail.

> What is this story about? It may be about the power of love to change one's mind or simply about the changing of mind. It may reveal that real emotions come from particularities and details, she saw how he suffered and how often he went to the privy. It may be about the power of human prayer. . . . It may be a tale of mitigating circumstances: we *ought to* take this action, but we *are forced to take* the second action.[95]

On this last point, remember that Cover observed that stories both articulate norms and gesture toward subverting them. This thus leads Cutter to conclude that "the story of Rabbi Judah's death should, then, not be the *locus classicus* for a particular action or decision, but a model for the uses of narrative. Like all good stories, it makes explicit the actual experience of the lived life."[96] Ambivalences, ambiguities, anxieties—these pervade life and are often expressed in good, rich, stories. As it is good for life to emulate (certain) stories, it is no less true that (certain) stories embody incarnated life.

Another ardent contributor to narrative ethics is Laurie Zoloth (1950–). She instructs time and again that narratives are necessary for Jewish ethics, but in and of themselves they are insufficient: stories must be coupled with law for an ethical argument to be Jewishly grounded.[97] She demonstrates this claim in her arguments for reconceptualizing healthcare resource allocation in favor of a system in which encounters matter more than materials, relations more than things, responsibilities more than rights. "If we are to develop new language beyond individual entitlements," Zoloth writes, "it must be language rooted in story and community that draws from a method that is *itself* dialogical and communal."[98] This dialogical and communal method is Talmudic in nature, insofar as it describes and reinscribes the interpenetrating relationship between individual and collective just as law and narrative do in the Talmudic corpus itself, and in so doing "creates the social and shared ground for justice."[99]

Zoloth turns to the Book of Ruth as the locus for her meditations on this renewed and just healthcare system. She contends that the "story is meant by the author of the text to be held in tension with the other narratives [of the Bible], and this tension is part of the methodology of temporal simultaneity of the system itself."[100] That is, just as any one particular healthcare moment exists simultaneously alongside myriad other encounters, so too this one story resides and resonates in juxtaposition to other stories. Such coupling Zoloth cherishes as she delves deep into the complex relationship between Ruth and Naomi, her mother-in-law. Through careful analyses of the biblical text and relevant commentaries, Zoloth traces how justice and reparation emerge from "the move not only from famine to feast, but also from enslavement to service and from exile and displacement to community and place."[101]

Certainly the Book of Ruth is not the sole source of language of responsibility, nor is it the only text outlining minimal standards of care for the impoverished.

> [Rather,] it does better than the other texts in providing a basis for talk about health care as a social, daily experience that will define the nature of community in major ways. While it acknowledges scarcity, it transforms its power: scarcity, the famine that lurks beneath every abundance, does not threaten the fabric of human community if the demand of relationship is consistently heard. It is not a text about violence; it is a text about the experience of aging, vulnerability, and solidarity in the face of death. The Ruth text deals with the most difficult issue of agency, that of the limits set between the self and the *ger*, the other, the stranger in the land.[102]

Its grounding in women's experience and its demonstration of "the power of women's choices to remake the world" make it a necessary source for any comprehensive rethinking or ethical deliberation of healthcare delivery.[103] Whereas other texts may direct what to do, this text suggests; it opens new vistas that other ways of reading and thinking could not even imagine.

Essentially . . .

Newman is correct to observe that all Jewish ethicists and bioethicists—formalists, covenantalists, narrativists—point at stories or at pieces of stories in some way or another within their work. Yet, "if all agree that doing Jewish ethics necessarily entails reading and interpreting Jewish texts, there is anything but agreement on which texts to read, or how."[104] The desire of some scholars to have all Jewish ethicists and bioethicists read the same texts in the same way may be counterproductive, however. Such a strong agreement on which sources matter and how they are to be interpreted would necessarily snuff out communal deliberations on perplexing problems; disagreements for the sake of heaven would dissolve into mute consensus and rote obedience. The health of the field of Jewish bioethics in particular and ethics more generally revolves around the necessary tasks of reading and interpreting, of creatively linking ancient sources with modern circumstances. As Harvey Fox says, "this process of exegesis and interpretation is the most important device that Jewish tradition used in order to be able to stand simultaneously in the classical tradition of Judaism and in the contemporary world."[105] The very vitality of Jewish bioethics and ethics rests in part on the practice of reading—and essentializing—stories.

In addition to the concerns adumbrated above, essentializing a story has a further ramification. It highlights selected bits and ascribes them greater import than the rest of the bits comprising a story. In this way essentializing allocates to the chosen pieces some level of normative stature that they otherwise may not deserve. On the other hand, it could very well be that within the story itself the chosen pieces are indeed morally or legally significant. But as we shall see here in regard to the Chananya story, the bits selected and highlighted by bioethicists are of questionable moral, legal, and even bioethical, significance. Scholars' foreshortened readings impose upon narratives their agendas, and these reading strategies merit interrogation.

For example, finding support for a particular position is a common goal among bioethicists. Achieving this goal therefore shapes how one reads classic sources. It constrains what one will see as relevant in a particular story, and, further, it will construe what appears as relevant as something that it may not be. As Wimpfheimer demonstrates, a goal-oriented reader of a narrative may find something in it that it may or may not contain.[106] Goal-seeking reading strategies, especially if that goal pursues (supporting) norms, invariably flattens a narrative into a norm and strips it of its peculiarity. This simplifies the story, eviscerating it of its time dimension, complexity, character development, denouement (if there is one), and all the other elements that make stories so interesting and compelling. To be sure, these other elements are part and parcel to each and every bioethical dilemma in clinical settings. Since there is no generic bioethical case in reality, why should the stories that undergird bioethical deliberation be denuded of their particulars and rendered generic according to this or that author's agenda?

Perhaps there are other ways bioethicists can read stories that protect the integrity of the story—just as they would strive to honor and protect the integrity of the whole story of a modern patient. This book explores such possible ways. It may be, however, that reading texts—be they law or narrative or health report, for that matter—is necessarily a selective and interpretive exercise. Hermeneutics is inevitable. Even if this is true, which I suspect it is, there may be ways of reading that are more harmful than others, and ways that are more healthy than others—both in regard to the sources invoked and audiences intended. Before we get to those suggestions for vivacious reading and writing, however, we must traverse a particular story of death.

NARRATIVES, END OF LIFE CARE, AND COMPLICATIONS

The knot of narratives and Jewish bioethics is even more tangled when we look at the discourse concerning care at the end of life. Before teasing apart the various threads of this knot, we must first justify why this end of the cord of life merits so much attention here and now. Why not focus on another biomedical moment, like the beginning of life, or what constitutes the very length of life—the minutiae of daily choices like diet and exercise and sexuality and risk-taking behaviors? Why so much concern about our terminal end? Pressing sociological, emotional, and methodological reasons encourage this focus. The massive baby-boomer generation is no longer a baby. Nor is it even middle aged. People comprising this generation edge ever closer to geriatric care. Since they are the largest segment of the population in the United States, it behooves bioethicists to be concerned with their care, their dying, and their deaths. In addition to this human dimension, there is a humane dimension as well. Senescence is emotionally fraught for those experiencing it, as are dying and death itself. One's own death is troubling, and being involved—even tangentially—in someone else's death is also distressing. And methodologically, as has been alluded above, bioethicists need to be careful about both what they encourage others to do and how they encourage them. For these reasons, any clarity that can be generated regarding how to navigate the fraying ends of life will be particularly useful for patients, families, care providers, and rabbis.

Especially since the mid-1970s when Karen Ann Quinlan surprisingly survived long after the ventilator had been disconnected, Jews have increasingly bent themselves to the task of thinking through the moral morass surrounding intervening in someone's demise. Up to this point scholars resorted primarily to the massive books on medical ethics by Preuss and Jakobovits; Emanuel Rackman's 1956 discussion of this topic is void of any reference to classic texts, and although Fred Rosner's 1967 piece does wrestle with some sources, he ultimately defers to Jakobovits. The emergence of Harvard's notion of brain death in the late 1960s and early 1970s sparked among Jews vigorous debate about what constitutes death and, for that matter, dying as well as euthanasia generally.[107] Recent advances in pain management technologies and shifting opinions about "death with dignity" in the United States and Europe continue to inspire Jews to pay attention to various dimensions of euthanasia. Demographics also adds fuel to the urgency on this topic as more and more Jews find themselves living longer and wondering not when they will die but how, and more and more Jews exist in the "sandwich generation" in which they attend to the care of elderly and dying parents as well as for their own children.

For all these reasons and more, scholars, rabbis, physicians, and interested lay people have all added to the discussion of euthanasia in recent years. Whatever semblance of modest agreement may have existed among Jewish bioethicists regarding euthanasia in the earlier decades of the twentieth century certainly no longer is visible. The debate has fractured and fissured, in part because of the diversity of contributors to the discussion and in part because of the myriad ways these contributors go about reasoning to their conclusions.

On the other hand, an emerging trend throughout the contemporary discussion on euthanasia is the turn to a particular narrative—the story of Chananya ben Teradyon's final moments. This story (and the one of Rabbi Judah HaNasi and his handmaid to a lesser degree) serves as the pivot for most if not all euthanasia tracts.[108] Regarding this point Len Sharzer says:

> It is a universally accepted principle in Jewish bioethics that it is not permissible to hasten a death, but it is permissible to remove an impediment to the dying process. Nearly every discussion that enunciates this principle cites two stories from the Talmud, the story of the execution of Rabbi Chanina ben Teradyon by the Romans and the story of the death of Rabbi Yehudah HaNasi. Although most citations note that *halakhah* (normative standards of behavior) cannot be derived from *aggadah* (narrative), the stories are often said to support and, one might argue, validate the principle that although it is not permissible to hasten death (actively), it is permissible to remove an impediment to death.[109]

Sharzer's assumption that this principle is universally accepted may be contestable especially in the last few decades, but this is not my issue. What I want to focus on is that there are but two stories—and one of them in particular—around which contemporary bioethical deliberations about euthanasia turn. And I want to highlight Sharzer's observation that these stories serve to *support* and *validate* the principle of not hastening death but of removing impediments thereto. Sharzer echoes Hilde Lendemann Nelson's claim that bioethicists often invoke stories to reinforce certain claims or positions: Jewish bioethicists are no exception. What I will show is that how they go about pointing to these archetypal stories and invoking them for the tradition's imprimatur for whatever position they hold is problematic on many levels—for the story, for their arguments, for their audiences. Even though this story regarding Chananya's fiery end may be the most cited in Jewish bioethical discourse on euthanasia, it is underappreciated for all its complexity, depth, and diversity.

For example, some bioethicists contend that any and all euthanasia strategies are impermissible, and this story endorses that position. Others aver that passive euthanasia is palatable, and they say the story supports this claim. And a few argue that the story countenances even active euthanasia. Such disparate normative conclusions emerging from this single story is fascinating and disturbing. It suggests that the story is more complicated or ambiguous than what these scholars say it says—which itself should inspire further study of it. On the other hand it also means that the story is ambivalent insofar as it lends itself to dramatically apposite if not opposite strongly held attitudes. So what one scholar insists the story holds may be as true as what another scholar, advocating a different conclusion, champions. If both attitudes truly exist within the narrative, which should trump—which should form or inform a normative conclusion? No scholar, as far as I have found, has demonstrated a non-question-begging meta-ethical standard justifying the superiority of his or her extraction over others'.

Rather than get distracted with such a competition, I propose approaching this story differently. Recall in this chapter's introduction Nelson's outline of five ways we engage stories: we (1) read them, (2) tell them, (3) compare them, (4) analyze them, and (5) invoke them. As has already been shown, Jewish bioethicists most often invoke stories, pointing to them to bolster their arguments. Some dig a bit deeper and compare stories, teasing out certain details, juxtaposing elements as would be done in classical casuistry. Narrative ethicists would perhaps prefer the first strategy: reading stories so as to let the details and moral richness wash over the reader and soak in, so that the reader would imbibe the story's moral sensibilities. What I propose here is a more synthetic approach that (1) takes the story seriously; (2) tells stories that reinforce our moral complexity; (3) compares like stories across the textual tradition and thereby appreciates the polysemy, ambivalence, and ambiguity the tradition ensconces in its pages; (4) dissects the details of a story for its internal workings and how it prepares readers to receive what comes next; and still (5) invokes them as the "moral of the story"—warning, goading, precedential, and so on. The rest of the book endeavors to model this kind of thick reading of narrative.

This ambitious plan is beset, however, by a significant challenge. Time. It takes time to read, tell, compare, analyze, and invoke stories. Newman rightfully points out that "our moral lives are lived one decision at a time"—and each decision is itself a micro-story.[110] Our lives are nothing but stories leading to other stories, narratives built upon narratives. But, as Cutter warns, "the world of decision making does not wait for narratively inclined people."[111] It will take some time to tell this story and extract insights from it. Nevertheless, there is something urgent and burning about this project: from the fiery protagonist of the central story, to the scholars' sparks of certainty that theirs is the proper way to read and appreciate the story, and ultimately to the conflicted and dying patients whose lives gutter like sputtering flames. For these reasons and more, we must push on into the story itself.

Chapter 3

A Dying Story: Told and Retold

There has never been a death more foretold.[1]

Introduction

Understanding the interrelationships between narratives and norms in contemporary Jewish bioethics requires at least two projects, one that looks at what the stories are and one that examines how they are read. First, effort must be given to identify and analyze narratives in their textual contexts. This involves reading ancient stories as they are found in the textual tradition. Care should be given to locating and analyzing these stories in complementary textual witnesses. Second, how modern Jewish bioethicists read these stories requires scrutiny. Of course bioethicists are not the only scholars who turn to these stories for insight and guidance. For this reason it is important to juxtapose modern bioethical readings of these ancient stories with other modern interpretations. The first step appreciates the beauty and complexity of the stories in their own light, and the second highlights the difficulty these very stories pose for contemporary scholars. This chapter attempts to accomplish the first of these tasks; the subsequent two chapters fulfill the second.

There are hundreds if not thousands of stories embedded in classic Jewish texts. It would be impossible to survey them all and difficult to do them scholarly justice here. For better and for worse, in the field of modern bioethics only a handful receives attention of the modern mind. Curiously, just a scant few are cited in certain bioethical discourses. In the modern conversation on euthanasia only a couple of stories appear repeatedly. One regards Judah HaNasi who was dying of dysentery. When his handmaid witnessed his inconsolable discomfort, she seemingly prayed for his life to end, though his students prayed for him to be miraculously healed.[2] A second story also relates prayer and death. This time an old woman seeks escape from her increasing incapacitation. Rabbi Yose ben Halafta encourages her to stop her daily pilgrimages to the synagogue for three days, which she did and then immediately died.[3] A third story refers to 400 boys and girls captured and shipped off for sexual impropriety. Upon divining their future the girls jumped into the sea, which inspired the boys to leap to their deaths as well.[4] While narratively intriguing, such stories fail to capture contemporary bioethicists' imaginations. With the exception of the first about Judah HaNasi, most stories are usually mentioned only in passing in euthanasia literature.

There is, however, one story that stands out as exceptional—in importance and complexity. This one is viewed repeatedly if not unanimously as the *locus classicus* for this bioethical and lethal debate about euthanasia. Indeed, only this one serves as the pivot around which scholars turn their positions to either condemn or condone certain forms of euthanasia. Its pervasiveness within and critical contribution to modern bioethical discourse on euthanasia thus places it in a unique position meriting special attention here. In brief, it is a story about a particular rabbi's immolation by Romans. Our two-fold task is therefore to understand his flaming death as it is actually portrayed in the textual tradition, which is done in this chapter, and then to assess how contemporary scholars and bioethicists read, understand, and interpret the narrative, which follows in subsequent chapters.

The Context

The story central to this project is found in the Babylonian Talmud (hereafter BT), *Avodah Zarah* 18a, a tractate ostensibly concerned about issues of idolatry. Our *sugya* (story) resides in the first chapter of this tractate, the topic of which is the nature of the interactions between Jews and gentiles on gentile holy days. The Mishnah, the earliest layer of rabbinic law and probably redacted around 200 CE, simultaneously prohibits Jews from aiding and abetting gentiles in executing capital punishment and yet permits Jews to assist gentiles in constructing altars and public baths.[5] The *sugya* is located in the Gemara's discussion of the prohibition to assist gentiles constructing basilica, scaffolds, stadiums, or platforms.[6] Though scholars disagree on the precise date, it is commonly understood that the Babylonian

Talmud, comprised of both the Mishnah and Gemara, was redacted sometimes in the sixth or seventh centuries CE.

The main protagonist is R. Chananya ben Teradyon (hereafter, Chananya), a second-century Palestinian Tannaitic sage. Married with at least two daughters and two sons, Chananya was a well-respected scholar of his day. Before turning to how he died—the part of the story that most excites contemporary bioethicists—it is important to understand how he got to that moment. This summation follows the popular printed Vilna edition of the Talmud. This point is crucial because while most scholars today consult the edition of the Talmud that was printed in Vilna in the latter part of the nineteenth century, several handwritten manuscripts of the Talmud of this section that predate the Vilna edition also exist. In fact, of the 28 extant manuscripts of this tractate, five manuscripts still have this particular narrative in their pages. These manuscripts are: Paris 1337; Munich 95; Pesaro 1515; JTS Rab 15; and Jerusalem-Schocken Institute for Jewish Research 3654.[7] In retelling this story as comprehensively as possible, I incorporate these six competing manuscripts (the printed Vilna edition of course must be included) especially for the part of the story that most concerns contemporary bioethicists.

Two generations after the destruction of the Second Temple in 70 CE, the oppressive Roman Hadrianic regime (117–138 CE) in Palestine outlawed Torah study, Jewish religious observance, and rabbinic ordination (among other things), and often rounded up sages who flaunted these strictures and executed them in a public manner so as to assert its authority. According to the Talmud, Chananya was arrested alongside his colleague R. Eleazar ben Perata. In their discussion about the accusations against them, Eleazar praises Chananya for being arrested for only one charge, that of studying Torah. Chananya demurs, praising Eleazar for being charged with five accounts and especially for studying Torah and doing *gemilut chasadim* (acts of good will). When Eleazar was brought before the Roman tribunal, he denied all accusations and was ultimately released without charge.

By contrast, when Chananya replied to the charge of studying Torah, he stated, "Thus Adonai my God commanded me." He was immediately sentenced to burning, his wife to decapitation, and their daughter consigned to a brothel tent.[8] According to the Talmudic rabbis, Chananya's crime that merited death was speaking God's name in public. When Chananya, his wife, and their daughter emerged from the tribunal, all three exclaimed the justness of the verdicts.[9] Chananya recited, "*The Rock! [God's] deeds are perfect, all [God's] ways are just*" (Deuteronomy 32:4a). (Given that his crime was speaking God's name in public, it is not unimportant that the prior verse reads, "*For the name of Adonai I proclaim, give glory to God!*" [Deuteronomy 32:3]).

His wife recited the latter part of Chananya's verse, "*A faithful God, never false, true and upright [is God]*" (Deuteronomy 32:4b). In this way she also announces that despite—or perhaps because of—this judgment against her and her family, God remains fair and just. Her crime, the Talmud declares, was not protesting against Chananya: she did not prevent him from speaking God's name publicly.[10]

The daughter recited, "*Wondrous in purpose and mighty in deed*" (Jeremiah 32:19a). The last part of the verse she left unsaid reads, "*Whose eyes observe all the ways of men, so as to repay every man according to his ways, and with the proper fruit of his deeds*" (Jeremiah 32:19b). Rashi interpreted her verse to mean that God pays attention even to the way a woman shapes her gait, suggesting that she believed that God observes human behavior and repays humans even for such trivial behavior as foot placement.[11] The Talmudic rabbis support Rashi's interpretation; for them, she is guilty of immodesty and thus merits punishment.[12] A perhaps more cynical daughter would have recited the prior verse to the one actually uttered: "*You [God] show kindness to the thousandth generation, but visit the guilt of the fathers upon the children after them. O great and mighty God whose name is Adonai Tzva'ot [The Lord of Hosts]*" (Jeremiah 32:18). In this instance the daughter's biblical incantation would not be about God's justness as much as about the incomprehensibility of punishing children for parental crimes, even though this tenet is ensconced in the Decalogue.[13] Indeed, instead of Jeremiah's verse she could very well have invoked Ezekiel's position, "The person who sins, he alone shall die. A child shall not share the burden of a parent's guilt; the righteousness of the righteous shall be accounted to him alone, and the wickedness of the wicked shall be accounted to him alone."[14] This would have expressed her theological endorsement of her father's punishment, yet rejected her own. However attractive this hypothetical response might be for a modern sensibility, the Talmud records the daughter's solidarity with her father and acceptance of her own fate.

The Talmud then presents another story about Chananya, this time visiting his sick colleague R. Jose ben Kisma. Jose reminded Chananya that even though Rome destroyed the Temple, slew sages, and dispersed Israelites into exile, Rome nonetheless had divine imprimatur to reign. Jose wondered why Chananya continued to study Torah, gather assemblies, and teach from a scroll in his lap—all activities prohibited by Rome. Chananya retorted that Heaven would show mercy. Jose remonstrated with uncanny prescience, "I will not be surprised if they do not burn you along with the Torah." Chananya then asked, "Where do I stand in regard to the World to Come?" Jose inquired what sort of deed he had done to deserve such a reward, and Chananya told him a time when he gave monies to the poor out of his own pocket instead of from the public coffers. Jose replied, "Would that your portion be my portion, your lot my lot."

It is here that the story central to our project appears. The footnotes point to manuscript variations.

THE STORY

1. They said that only a few days had passed since the passing[15] of R. Yosi ben Kisma when all the elite of Rome[16] went to bury him[17] and eulogize him with a great eulogy.[18]
2. Upon returning, they found R. Chananya ben Teradyon sitting and engaging[19] in Torah, convening substantial gatherings, and a Torah scroll kept on his lap.[20]
3. They brought him and wrapped him in the Torah scroll, and they placed piles of twigs around him, and lit them aflame.[21] They brought tufts of wool soaked in water, and placed them upon his heart so that his soul would not depart quickly.[22]
4. His daughter said to him, "Father,[23] to see you thus!"[24] He replied to her, "Were[25] I to be burned alone it would be a difficult matter for me. Now as it is me who burns and the Torah scroll is with me,[26] the one who will address the affront[27] to the Torah scroll will also address the affront to me."[28]
5. (a) His students said to him, "Teacher, what do you see?" He replied to them, "The parchment burns but the letters soar."[29] (b) "Also you can open your mouth and the fire will enter you."[30] "It is better that the one who takes it is the one who gave it. One should not injure oneself."[31]
6. The executioner[32] said to him, "Teacher, if I increase the flames[33] and remove the tufts of wool from your heart, will you bring me to the World to Come?" He said to him, "Yes." "Swear to me." He swore to him.[34] Thereupon he immediately increased[35] the flames and removed the tufts of wool from his heart. His soul quickly departed.[36] He (the executioner) even jumped and fell into the fire.[37]
7. A *Bat Kol* (a heavenly voice) came and said, "R. Chananya ben Teradyon and the executioner have been assigned to life in the World to Come."[38]

8. Rabbi (Judah HaNasi) cried and said, "Some acquire his [eternal] world in one moment, and some acquire his [eternal] world in many years."[39]

While this concludes our main *sugya,* it contains some curious inconsistencies. The larger context presents at least two versions of Chananya's arrest, trial, and punishment. In the first he is charged with studying Torah, for which he is taken to a tribunal that ultimately sentences him, his wife, and his daughter. But in the second there is no tribunal that ultimately sentences him. Rather he is apparently guilty of three interrelated crimes—studying Torah, gathering assemblies, and teaching from a scroll—for which only he and not his family is punished. Moreover, in the first story there is no mention that he was to be burned with a Torah scroll, while this is a prominent detail in the second.

These internal narrative inconsistencies are compounded when the manuscript traditions are taken into consideration alongside the popular printed Vilna edition. All manuscripts agree on the sequence of this central scene. The least contested elements among the manuscripts are when this incident occurs (paragraph 1 in the story above),[40] the statement of the *Bat Kol* (7), the cries of R. Judah HaNasi (8), and what Chananya first responds to his students about the parchment and letters (5a).

What Chananya says next to his students (5b) is most perplexing, however. All manuscripts agree that his students encourage him to open his mouth so the fire or flames can enter his mouth. They echo a larger Tannaitic notion that one condemned to death by fire ultimately expires when flames enter through the mouth.[41] Only one manuscript, Paris 1337, asserts that this action would bring him some dimension of comfort.

In response to his students' lethal suggestion Chananya offers a grand principle: "It is better that the one who takes it is the one who gave it. One should not injure oneself." The last sentence in Hebrew is: *v'al yechavel hu b'atzmo.* Four of the six existing manuscripts, including the Vilna edition, corroborate this statement. Not one but two manuscripts offer a dramatically different statement, however. According to these versions Chananya says to his students, "It is better that the one who takes it is the one who gave it. *I* will not injure *myself*" (*v'al echavel ani b'atzmi*).[42] This textual countertradition suggests that Chananya did not seek to teach his students a general rule about proper behavior in this circumstance but merely wanted to demonstrate his own self-restraint.[43] With his mouth shut against the flames, Chananya leaves open the possibility that others, if they found themselves similarly being burned with a Torah scroll and wet tufts of wool, could or perhaps should even open their mouths so the fire could asphyxiate them.

Such a significant departure by a third of all existing manuscripts from the popular printed version raises critical questions. We could inquire how it was that the Vilna edition came into being, and which manuscripts were consulted in its compilation. Such text-historical investigations merit attention but they are digressions here. Rather, the existence of this counter tradition points to a vital issue: is Chananya's statement articulating a rule or an exceptional personal commitment? How one answers this question becomes critical if not lethal in those contemporary bioethical arguments that focus on these dying words of Chananya.

There are still further details of this story that perplex. Take, for example, what Chananya says to his daughter (4). It should not be surprising that she was present at his death, as it was common practice for Roman executions to take place in public venues. It is unclear here which daughter this is, however: is it the one who was sentenced earlier to a brothel, or is it Beruriah who is mentioned in the *sugya* immediately following this one? Either way, she had such close contact with him in his dying moments, so close they could share profound emotions. She expressed what any child would say upon seeing a beloved parent in such a state of trauma. Ever her father, Chananya, calmly soothed her angst. Do not worry, child—he seemed to say—for God is with us, and is even with me in this terrible moment. This is no mere articulation of faith but a cross-generational transmission of values. That he tenderly encouraged his daughter to take solace in the fact that God will right this wrong, Chananya bequeathed to her a particular worldview, complete with its value system, in which she could take comfort. In a way, these words were his living will.

Chananya's students were also present (5). Though their emotions were not as distraught as his daughter's, their motivations and curiosity as to the nature of his experience were unclear. Ever his students, they continued to see this burning man as their teacher and they beseeched him for a final lesson. "What do you see?" they asked (5a), assuming a burning man could pause his pain so as to put words to his visions. Chananya's response that the parchment burns but the letters soar nonetheless expressed a powerful rebuttal to Rome. The regime could destroy the material aspects of God's revelation but they could not touch revelation itself. The very words of revelation escape and ascend back to their source, ready to be promulgated again.

On the one hand, Chananya suggested to his students to anticipate an onslaught by Rome, and on the other, he instilled in them the conviction that their faith transcends parchment, pyres, and temples, that their faith will ultimately prevail.

It should not surprise anyone that Chananya shared with his daughter one kind of final message and offered a completely different one to his students. To one he offers centering and solace even when he is in the throes of death, but to others he offers a lesson in political theology, a lesson that could well be offered at any moment in a teacher's life and need not be one's last. Could it be that the juxtaposition of these conversations highlights some underlying value of the *stam*, the anonymous redactor of the Talmud? The reader would expect a dying parent to share with a child some profound insights about the ways of the world and where comfort could be found. So when the reader next encounters students beside Chananya's burning body, it would not be unreasonable to anticipate that their conversation would be similarly intimate. Yet it is anything but intimate. Despite the whiff of the heartfelt conversation with his daughter still hanging in the air, Chananya's students have the gall to ask a seemingly tasteless question that betrays a stunning insensitivity. Yet not to disparage the students altogether, it could be that they so loved their teacher that they could not fathom that even this moment of searing flesh could not hold within it yet another kernel of truth or illuminating insight their teacher could burst open for them.

And yet the students were not done with their teacher. They then encouraged Chananya to avail himself even more to the lethal fingers of the flames by opening his mouth (5b). A few manuscripts, as noted above, add a word or two to couch the students' suggestion in terms of compassion: by opening his mouth he would hasten his death and bring him more comfort than he was currently experiencing. Yet none of the manuscripts could suppress or erase the broader political dynamic of the teacher–student relationship altogether. Specifically, however much deference these students were determined to show their beloved rabbinic teacher, they also stood to gain much upon their teacher's demise: some, or at least one, could ascend to his post. Would it be too farfetched or cynical to impute to these students some professional ambition?

And what can be made of the exchange between Chananya and his executioner (6)? For one reason or another, the executioner seems to be a fairly knowledgeable fellow; he's informed enough about his victim's theology to inquire about the possibility of accessing the World to Come. As noted above, he is not the only Roman who sought such eternal life.[44] What is the significance of a pagan Roman, a *quæstionarius* no less, seeking this peculiarly Jewish religious reward? Could this bit of the story be an expression of rabbinic supersessionism, a kind of triumphalism in which the rabbis assert that even the most cruel of the pagans cannot but be attracted to Judaism, even as they go about the bloody business of executing individual Jews? Had this conversation come before the prior two or was the only one to exist, it would have been reasonable to anticipate Chananya's rejection of the executioner's self-centered request. But this is not the case; no manuscript reorders the three conversations in any way. Rather, this is the last conversation Chananya has before he dies. We cannot ascertain precisely why Chananya agreed to bring his executioner into the World to Come, but we certainly do know he was successful: the *Bat Kol* announced it! The executioner's plan and performance received a divine seal of approval.

Taking these conversations seriously and sequentially, a strong theological trope appears. Chananya was caught promulgating Judaism in a public manner and for which he was burned alive with his Torah scroll. His verbal responses during his immolation expresses a trifold manifesto: (1) believe, child, for God exists and creatively exacts justice; (2) be steadfast, students, for God's revelation endures; (3) welcome, stranger, for God's bounty is available to all who seek it. Put more succinctly, Chananya's dying theology echoed that of Judaism writ large: God is the God of Creation, Revelation, and Redemption.

Such is a theological analysis of this central story. But, like William Cutter, we should not be content with reading a story once. We should return to it, this time with another tack that is less metaphysical and more physical. Consider again Chananya's response to his daughter. To her he insists that he is in a reasonable situation: nothing can or should be done by humans to intervene in his dying process. But to his students Chananya is explicit: neither you nor I will do anything to interfere in my dying process. And to the executioner, his opinion could not be more dramatically different: yes please, do something to intervene in my demise. Like the theological lens, this more fleshy perspective understands the three conversations occurred sequentially and not simultaneously. That is, time elapsed. And through time Chananya continued to experience his own demise, even as he had these conversations seriatim. If we can assume that Chananya was just as human as the rest of us, it would be inhuman and unreasonable to assume he was impervious to the flames throughout all these conversations. Given this fact, why would he not change his opinion about human agency generally and intervention in his dying in particular? To impose upon him consistency throughout his dying moments would amount to removing from him his very humanity, not to mention denying his statements explicitly expressed in the *sugya*. We return to this more physical perspective below.

DYING AGAIN

Chananya does not die once in rabbinic literature, however. He dies several times; four times in total. In addition to the narrative of Chananya's demise in *Avodah Zarah,* Chananya dies in three other rabbinic sources. Indeed, it could be true that some of these sources predate the story found in the popular *Avodah Zarah,* which raises additional questions about which Talmudic manuscript if any represents what actually happened. But before turning to such historiographical issues, we should examine these competing stories of Chananya's dying moments. As will be demonstrated, they do not agree who was present when Chananya died, how he died, or what he said to whom.

Sifre Devarim §307

The oldest of these stories is found in a third-century collection of *halakhic midrashim* (legal stories) on Deuteronomy.[45] The characters populating these stories are all Tannaim (approximately 10–220 CE), suggesting that they are at least as old as the Mishnah even if the collection was not redacted until later. This particular *piska* (interpretation) comments on Deuteronomy 32:4a, "*The Rock, [God's] deeds are perfect.*" The first set of interpretations reinforces the plain meaning of the verse that as God has no flaw so too is creation and God's judgment thereupon. The interpretation immediately preceding our story ends with a person declaring that God's judgment upon him was fair.

1. Another comment on "*The Rock, [God's] deeds are perfect.*"
2. Chananya was apprehended and condemned to burn with his scroll.[46]
3. When he was informed of this, Chananya recited, "*The Rock, [God's] deeds are perfect.*"[47]
4. His wife was then told, "Your husband has been condemned to burn[48] and you are to be killed," to which she responded,

"*A faithful God, never false, true and upright is [God]*" (Deuteronomy 32:4b).

5. They informed his daughter, "Your father is to be burned and your mother killed, and you will do labor,[49]" and she responded with "*Wondrous in purpose and mighty in deed, whose eyes are open*" (Jeremiah 32:19a).
6. R. Judah HaNasi said, "How great were these righteous persons,[50] for in the moment of their trouble they invoked three verses about the righteousness of judgment, which are unequaled in all Scripture. The three of them intended their hearts and justified the judgment upon them."
7. A philosopher (*philosophos*) stood up and spoke to the [Roman] governor,[51] "My master, do not boast[52] that you have burned the Torah, for the place to which it has returned is its father's house."[53] The governor replied, "Your judgment will be the same as theirs." The philosopher said, "Those tidings given me are good,[54] for tomorrow my portion will be with them in the World to Come."[55]

This *Sifre* version of Chananya's demise focuses on his—and his family's—incredible righteousness, even equanimity, in the face of a seemingly baseless decree. Indeed, it is unclear precisely what crime he committed that warranted his and his wife's deaths and his daughter's consignment to do "labor." Yet this version echoes the Talmudic story of the three convicted individuals reciting biblical verses, the same ones in fact, when they hear their sentences (paragraphs 3–5 above). Though R. Judah HaNasi does not cry in this story as he does in the Talmudic account, he nonetheless offers a concluding moral (6). There in the Talmud he laments the apparent injustice that a lifetime of struggle and an instant's enlightenment can enjoy the same dessert of eternal life, here he extols those who do not challenge but rather justify the lethal judgments against them. Put differently, in the Talmud R. Judah HaNasi cries over just desserts and in the *Sifre* he praises desserts that are justified.

The exchange between the philosopher and the governor (7) seems to stand in lieu of the conversation between the executioner and Chananya as found in the Talmud.[56] It is intriguing that the philosopher remonstrates the governor for celebrating the burning of the Torah; however, he does not challenge the justness of this act or the fact that Chananya was also burned to death in the process—that would undermine the purpose of the whole *piska*! Indeed, we do not know whether Chananya was actually killed, only that he was scheduled to die. Nonetheless, the philosopher makes a theological rather than purely logical argument: no worldly fire, and certainly no government, can destroy the Torah altogether. For burning the Torah merely returns it to its origin and in so doing returns revelation to its source, a source that is always ready to send it forth again. This point obviously parallels the one made in the Talmud when Chananya told his students that the parchment burned yet the letters soared high. The government may have killed a man and burned his book, but the government failed to squelch what both represent. So when the governor informs the philosopher that his effrontery earned him the same fate, it is not surprising that the philosopher speaks jubilantly and in theological terms about acquiring a portion of the World to Come.

Furthermore, it is uncertain whether the Torah scroll was actually burned when Chananya burned, if he was at all. Whereas it was initially decreed that he would be burned with his scroll and he was informed of this, the Torah scroll slips from the narrative until the philosopher speaks up. Curiously, it is through the philosopher's mouth that Chananya is equated with the Torah scroll itself. Chananya's wife and daughter are informed that he is to be burned, but they are not told that his scroll would be consumed as well. And R. Judah HaNasi, neither here nor in the Talmudic version above, bemoans the fact that the sacred Torah scroll was burned. Ambiguity on this detail remains.

What can be made of time in this version of Chananya's end? Whereas in the Talmudic narrative duration cannot be denied, here it is unclear if any time passes. Indeed, it could very well be possible that while Chananya and his family extol the justness of the judgments imposed upon them on one side of a courtroom door, the philosopher accosts the governor on the other side. Only Judah HaNasi's comments seem to be sequentially later. Like a tesseract, diachrony seemingly folds into synchrony.

BT Kallah 51b

The next two versions of Chananya's death exist in what are called the Minor Tractates. These are sections of the Babylonian Talmud for which there is no matching Mishnah. Not only do they lack that prior textual foundation, they are considered post-Talmudic in provenance; their origin could be as late as 1100 CE, or as early as the third century CE.[57] For these and other reasons, these Minor Tractates, are often considered less important than the rest of the Talmud.

The first, version in *Kallah,* recalls an incident when Chananya mixed money to be given to the poor with his own funds.[58]

1. He sat in confusion,[59] saying, "Woe is me! Perhaps because of this I am obliged to die[60] for Heaven."
2. While he was still in this troubled state, an executioner[61] came and said to him, "It has been declared against you that you are to be wrapped in your Torah and burned,[62] and Israel with you."[63] He stood and they wrapped him in his Torah and surrounded him with bales of sticks.
3. When the light peeked at him, the light retracted and distanced itself from him.[64]
4. The executioner arose in amazement and asked him, "Rabbi, are you the one that is decreed to burn?" "Yes." "Why has the light extinguished?" "I swore to it by the name of my fashioner that it not touch me until I know if the decree against me is from Heaven. Give me a bit of time and I will inform you."
5. The executioner sat in confusion, saying, "Why should those who decree death and life upon themselves suffer the yoke of [earthly] government?"
6. He then said to Chananya, "Arise, go—and whatever the government wants to do to me they will do."
7. "Fool!"[65] Chananya replied. "It is confirmed that the decree against me is from Heaven, and if you do not kill me, God has many killers—many bears, leopards, lions, wolves, many serpents and scorpions[66]—to attack me. Ultimately, God will exact lethal punishment from you for my blood."[67]
8. The executioner then understood the situation and immediately stood and fell on his face. As he perished, he made his voice heard from within the fire,[68] saying, "*As you die, I will die and there I will be buried* (Ruth 1:17). As you live, I will live."
9. Immediately a *Bat Kol* came forth and said, "R. Chananya ben Teradyon and the executioner have been assigned to the World to Come."[69]

This *Kallah* narrative clearly stipulates Chananya's crime, and Chananya acknowledges that the punishment for it is death. But whereas there was a courtroom scene in the *Sifre* story, it is absent here. Rather, the story continues like the *Avodah Zarah* version with Chananya immediately brought out for a public burning. But it remains patently unclear why the Roman regime cares about the fact that Chananya fudged charity monies that were to be distributed primarily but not exclusively among Jews, or why it would do the dirty business of executing Chananya for this transgression linked explicitly to a Jewish festival. Nonetheless, the punishment is familiar: Chananya is to be wrapped in his scroll and burnt. And conversing with the executioner is also a familiar trope. Yet absent here are Chananya's family, his students, the philosopher and governor, and Rabbi Judah HaNasi, along with their scriptural recitations, inquiries for lessons, protestations, laments, and praises.

Nonetheless, the characters included here merit a closer scrutiny. The executioner in this story appears to be even more conscientious than the one found in *Avodah Zarah*. Whereas in that popular version the executioner schemes with Chananya for a goal he desires (life in the World to Come), here the executioner offers to assume whatever punishment the government will mete out for the captive's escape.[70] There Chananya acceded to the executioner's selfish plan; but here Chananya disparages this executioner's selfless plan. He knows full well that one way or another, his death is inevitable. The executioner comes around to understand that the Jewish God's justice cannot be otherwise, but, it seems, he cannot fathom much less tolerate operating under the seemingly arbitrary Roman justice system, and so he jumps to his flaming death alongside Chananya. His dying voice cries out his intention to live, die, and be buried next to Chananya—an intention that secures divine imprimatur.

Whereas time seems to have collapsed in the *Sifre* version, time is explicitly mentioned here. Indeed, Chananya requests some time to figure out whether his punishment is divinely wrought. While Chananya awaits divine insight, the executioner meditates upon a perplexing question and discerns an illuminating observation. These simultaneous parallel revelations—one from above, the other from within—occur in time and through experience. One occurs from within the flames that do not burn, the other happens outside or next to the flames that do not consume. Could it be that proximity to this unnatural fire sparked each man's revelations? Could this be a fairly obvious reference to the unconsumed burning bush Moses encounters (Exodus 3:2–6)? Like Moses, both men were transformed through their experiences. The story explicitly states that each man sat in confusion, but in time and through exposure to each other and the bizarre fire each man achieved clarity.

This dual transformation is but a singular gesture: the confirmation of faith in a God that keeps promises. From Chananya's perspective, God would ensure that the death he merited for his religious transgression would be achieved either now or later. The executioner saw in this Godly justice a consistency the Roman government he served could not enact; he knew the government would slay him for letting Chananya go. Whereas his unjust death would perversely satisfy the regime, Chananya's and only Chananya's death would satisfy God. Perhaps it was this insight that inspired his conversion—and I say conversion because borrowing Ruth's statement has long been acknowledged as the quintessential expression of a proselyte.[71] As in *Avodah Zarah* the executioner's explicit conversion elicits divine endorsement and reward, which, unlike *Avodah Zarah*, this selfless executioner either did not know about or did not overtly seek.

BT Semachot 8.11

The last version of Chananya's demise, also in the Minor Tractates, is found in the section on mourning, euphemistically called *Semachot* (happiness).[72] This particular story concludes a litany of stories of other rabbi's brutal executions by the Romans.

1. When R. Chananya ben Teradyon was arrested[73] for heresy,[74] they sentenced him to be burnt,[75] and his wife to be killed, and his daughter to sit on a pile.[76]
2. He (Chananya) said to them, "What have they decreed against that poor woman?" They replied to him, "She is to be killed."[77] He exclaimed this verse upon her, "*God is just in all [God's] ways*" (Psalms 145:17).[78]
3. She said to them, "What have they decreed against him, my Teacher?" They said to her, "He is to be burnt." She exclaimed this verse upon him, "*Wondrous in purpose and mighty in deed*" (Jeremiah 32:19).[79]
4. And when they burned him they wrapped him in a Torah scroll and burned it.[80]
5. And his daughter cried out, wept, and wailed before him.[81] He said to her, "My daughter, if about me you cry, if about me you weep, it would be better that the fire consuming me be fanned but not a fire that is not fanned,[82] as it is said, '*a fire fanned by no man will consume him*'" (Job 20:26).[83]
6. [He continued,[84]] "But if about the Torah scroll you wail, behold, the Torah is fire and no fire can consume fire.[85] Behold, the letters soar. But the fire consumes only the skin."[86]
7. "For even great servants of the king are punished by inferior servants, [as it is stated,] '*That is why I have hewn down the prophets, have slain them with the words of My mouth*' " (Hosea 6:5).[87]

It is patently clear here that Chananya's crime was heresy (paragraph 1 above), though it remains a mystery as to what precisely he did. Heresy, however, differs from blasphemy, which is what Chananya was guilty of in *Avodah Zarah* where he spoke God's name aloud in public. Heresy is the expression of theology contrary to what is acceptable.

And according to Maimonides, heretics do not have a portion in the World to Come; indeed, they should be cut off from the community because they claim one or all of the following: (a) there is no God and the world is without a leader; (b) there are more than one leader (that is, gods) for the world; (c) God has a body; (d) God was not alone when creating the world; and (e) stars are intermediary powers.[88] Insofar as many of these tenets reflect Roman—and other pagan and Christian—theologies, why would the Roman government care to execute Chananya for heresy, for being attracted to Roman ways and whys of worship? Would not heresy be an issue internal to the Jewish community? Moreover, even according to Jewish law it is uncertain whether heretics warranted death. The only thing that is certain is that they do not gain eternal life in the World to Come; rather, they are to suffer eternally in Gehinnom.[89]

Akin to the popular Talmudic story, this narrative also conveys that Chananya's family was punished alongside him. Chananya expresses great concern for one who must be his wife, when he inquires what "that poor woman" is to suffer (2). Depending on the manuscript of this text, Chananya recites scripture extolling divine justice. Such verses express the opinion that it is proper that his wife should die because of his heresy.

Given his concern for his wife, it would be natural to expect that the woman speaking after him was his wife (3). Yet this conclusion may be mistaken. First, this woman does not call Chananya her husband but rather *rabbi*, my Teacher. This appellation could easily be spoken by either his wife or his daughter. And what does she say? She invokes a verse from Jeremiah that explicitly articulates the faith that God's justice exacts just desserts: each gets what each deserves. Again, such an opinion could understandably come from the mouth of either his wife or his daughter. Yet this is what Chananya's daughter—not wife—says in the *Avodah Zarah* version of things. And, as we did above, if we take into consideration the immediately prior verse that reads, "*You show kindness to the thousandth generation, but visit the guilt of the fathers upon their children after them. O great and mighty God whose name is Adonai Tzva'ot [Lord of Hosts]*" (Jeremiah 32:18)—a strong argument is plausible that his daughter is the one speaking here. Transgenerational punishment would be a more personal concern for his daughter than for his wife.

Indeed, Chananya articulates the moral of his situation (5) to his daughter and not to his wife. The moral—"it would be better that the fire consuming me be fanned but not a fire that is not fanned"—merits contemplation, however. To buttress the moral, Chananya cites

Zophar the Naamithite who insists to Job that the wicked will be devoured by a fire fanned by no mortal.[90] Rather, their fire would be fanned from no other place than Heaven. Since the wicked are to perish by divinely fanned flames, Chananya reasons it would be better for him to die in a fire fanned by humans, such as by the Romans, lest he be considered wicked and suffer heavenly fire. Roman fire can destroy his flesh just as it destroys the parchment of the Torah scroll wrapped around him.[91] But a humanly fueled fire cannot touch essences, whether it be his spirit or Torah's revelation.[92] He does not say that such a fire is an illusion, but that this fire's power is limited to the earthly domain. The searing reach of Rome's authority extends only to material things, not to things that truly matter. Like his final words to his daughter in *Avodah Zarah,* here Chananya impresses upon his daughter a worldview overflowing with theological enthusiasm and political cynicism. Take these values, child, he seems to say, and live with and by them.

This story appears much more straightforward than the others. Few distracting details exist in it. For example, it does not contain tufts of wool soaked in water or miraculously retracting flames. There are no intervening students or executioners or philosophers. Efforts to intervene on Chananya's, or anyone else's, behalf are also missing. Chananya appears steadfast in this story, certain that he merits the punishment he experiences. He confronts his daughter's anguish and encourages her to see things differently, though it is left unsaid whether she hears him or accepts his paternal wisdom. The story ends with no explicit conclusion to their exchange or any final words about his fiery demise. With no World to Come or *Bat Kol* included, we are left wondering whether the rabbis who composed or redacted this story liked it, whether they felt its message too raw or naïve, whether the exchange between father and daughter was too intimate—to receive eternal or heavenly endorsement.

DYING THE SAME DEATH?

It is indisputable that there are at least four versions of Chananya's demise in the Jewish textual tradition:

(A) *Avodah Zarah*
(B) *Sifre*
(C) *Kallah*
(D) *Semachot*

A critical question is whether they tell commensurable tales. Are they synoptic? Or are they writing different stories altogether? Does Chananya die just once or many times?

It is possible to find commonalities shared across these narratives. To illustrate, just as *Sifre* (B) describes a court case in which Chananya and his family members articulate certain biblical verses, a similar scene exists in *Avodah Zarah* (A). More telling, however, are the other details shared across the versions. Chananya was scheduled to die, and to die by burning—he was not to be strangled, stoned, or decapitated (A, B, C, D). The punishments his womenfolk were to receive (A, B, D). Non-Jews knew about and desired to obtain the World to Come (A, B, C). Revelation, as instantiated by the letters of the Torah, is humanly indestructible (A, B, D). In addition to these elements, all stories articulated profound theological or theo-political musings—none concludes with a legal ruling. We will return to this observation below.

Yet, despite these common details, it is difficult to claim confidently that these texts tell the same story. They disagree about the nature of Chananya's crime, whether he was guilty of teaching Torah publicly (A), of mixing *tzeddakah* funds (A, C), of heresy (D).[93] Some describe his family as pious and textually knowledgeable (A, B, D), though they are not unanimous about what each person says. Different people populate these stories: family members (A, B, D), an executioner (A, C), the philosopher and prefect (B), Chananya's students (A), Judah HaNasi (A, C). The stories identify different people present at his burning: daughter (A, D), students (A), executioner (A, C). Only two (A, C) mention sticks or twigs piled around him.

It is fairly clear that Chananya speaks even while his flesh burns (A, C, D), yet it is uncertain with whom he speaks, whether with his daughter (A, D), his students (A), the executioner (A, C), or all of them and the texts just failed to record all the conversations. Still, what he says to whom remains convoluted; take, for example, his insights about the letters of Torah soaring: is that said to his students (A) or to his daughter (D)? Does he accede to the executioner's plan to expedite his demise (A) or does he repudiate the executioner's plea to flee (C)? In two versions R. Judah HaNasi comments on Chananya's death: in both he praises the family's piety (A, C), but in only one (A) he laments the apparently unfair accrual of equal eternal reward for unequal piety. The executioner is ambiguously presented as selfish (A) and selfless (C).[94] And reference to the World to Come appears only when non-Jews are included in the narrative (A, B, C), but not when only Chananya's daughter is his counterpart (D). Curiously, the

World to Come is not linked to any of the womenfolk, even though their piety to God and love for Chananya are manifest (A, B, D).

And what of the Torah scroll itself? Just one version integrates the Torah scroll into Chananya's crime (A). Others include it in the pronouncement of his punishment (B, C), and one introduces it only when he is about to be burned (D). Chananya speaks of the Torah scroll burning (A, D) though not in every instance; the philosopher referred to it having been burned (C). And only in one is it specified that he was wrapped in it while he was burned (A). These versions of the story not only disagree when the Torah scroll becomes an important detail, but the textual witnesses of each story also are not unanimous on the particularities regarding it.

Of course we should also highlight significant singular details. Wet tufts of wool, for example, exist in only one version (A), as does a retreating fire (C). The offer to increase the flames is found nowhere else except in the Talmud (A). And in one version the fire itself is all but absent (B).

Comparison of Chananya's moral—the *mutav*—statements (A, D) reflects distinct concerns. In one, Chananya impresses upon his students that self-injury is impermissible, be it a general rule or just for himself (A). In the other he expresses preference to die by human flames than by divine fury (D). "Avoid increasing morbidity" contrasts with "succumb to earthly mortality." One pertains almost exclusively to personal issues whereas the other incorporates theo-political dimensions.

At a more abstract level and one that relates to bioethical questions, what do these stories say about intervening in a person's dying process? Only two texts explicitly wrestle with this issue (A, C).[95] On the one hand, intervention means hastening Chananya's demise (A), on the other it means liberating him from imminent death (C). In one intervention elicits Chananya's ambivalence (A), whereas in the other he rejects it outright and consistently so (C).

Such inconsistencies across these versions—not to mention the disagreements among the textual witnesses of each story—render it impossible to declare with any semblance of certainty what actually happened to Chananya in his dying moments. The plethora of contradictory details precludes cobbling together a unified story as well as discerning which details are facts and which fictions. We could dismiss this problem by claiming by fiat one version true and the others fictitious elaborations. For example, we could say that the oldest source (B, or is it C or D?) is for this reason true or at least the truest; all the others merely expand upon it.[96] But if this were the case, then why is it not ensconced in the more authoritative source (A)? Moreover, none of the texts adumbrated above is from Chananya's lifetime; they were all redacted much later—centuries later—after he died. Remember, he lived in the second century; the earliest text emerged at least a hundred or more years later.

Some might then argue that at least some portions of these stories reflect what actually happened to him, such as dying by fire. This claim would then face the conundrum of identifying which pieces those are. For example, if he died by fire, did he have tufts of wool attached to his heart (A) or not (B, C, D)? A corollary problem would then be understanding those untrue details: are they merely rhetorical embellishments, allegories, legends, or something else? If rhetorical, what import should they be given? If allegories, to what do they refer? If legends, from whence did they come and why? Choosing to ascribe historicity or authority or both to a source—or portions of one source—is obviously nothing more than a modern choice, a way of reading, a kind of hermeneutic. Such preferences do not and cannot establish the veracity of the events as portrayed in any chosen text—no matter how much a modern reader might protest otherwise.

NORMATIVE NARRATIVES?

Though the intriguing complexities of these four versions of Chananya's demise might overlap, this very fact resists the claim that any one version, or portion of one or some, is historically true. It also undermines claims that any one version or portion thereof is authoritative and has, or should have, normative suasion. The plurality of details, perspectives, and concerns makes it all but impossible to assert with certainty what "the" rabbinic viewpoint (of Chananya, of dying, of intervening, etc.) is. Isolating unambiguous rabbinic evaluation of intervening in a dying man's dying is a futile task, especially in Chananya's case. It would be better, says Barry Wimpfheimer, to read such narratives "not as attempts to communicate fixed cultural values or ideas but as vehicles for storytellers to process their cultural worlds."[97] Instead of flattening these narratives to a single norm, we should rather appreciate them in all their contradictory glory and glimpse the authors' struggles to make sense of a death and dying in complex circumstances.

On the other hand, if we were to take one source as authoritative, say *Avodah Zarah,* would it be possible to ascertain a clear norm therein? As will be seen in Chapter 5, many contemporary bioethicists do just this when making arguments for or against euthanasia. So as to better appreciate their hermeneutics, we return, qua Cutter, to *Avodah Zarah* as it is actually recorded in the Talmud for a final search for a norm or norms about intervening in someone's dying process.

The *sugya* as a whole articulates internal ambivalence about intervention, especially if we look to Chananya as the articulator of rabbinic values. Chananya's comforting words to his daughter bespeak a level of equanimity that prefers no human intervention. To his students, however, he adamantly refuses the proposal to be personally involved in either suicide (by opening his mouth for asphyxiation) or self-injury. And yet he entertains and even agrees to the executioner's plan to hasten his death; he welcomes intervention. Chananya expresses at least three strongly held opinions about whether he or anyone else should influence his dying process. His opinions cannot be collapsed into an overarching one without doing irreparable harm to one and all. And, as noted above, it cannot be ignored that he expresses these various opinions to different people at different moments in time. Pointing to one of his opinions as definitive of the whole *sugya* erases the temporal nature of dying generally and his dying in particular.

This ambivalence is further complicated by ambiguity.[98] It is unclear whether Chananya said to his students a general rule ("one should not...") or a personal commitment ("I will not...") others need not follow. And in regard to the executioner, the modern reader seeking a norm from Chananya faces a difficult problem. If the reader is smitten by the executioner's offer to remove the tufts of wool, it would mean that he or she finds the removal of life-sustaining treatments, or passive euthanasia, reasonable and supports it therefore. The reader could also focus on the executioner's proposal to increase the flames and thereby secure endorsement for proactively hastening someone's death, or active euthanasia. Or the reader could point to Chananya's assent to the executioner's plan and therefore promote the dying person's conscientious participation and thereby construe it as assisted suicide. Such ambiguity forces the modern reader seeking a norm to emphasize one element or another with seeming arbitrariness.

Despite content ambiguity and normative ambivalence, modern readers, especially Jewish bioethicists, nonetheless look to this Talmudic story—and not to others—for a definitive direction on what to do with dying people. Bioethicists are not alone among contemporary readers, however, who look for guidance in this story. As we will see in the next chapter, some modern scholars plumb this text for its theo-political themes and norms. Curiously neither these scholars nor the bioethicists unanimously claim a particular version of the story or details of it to be true. In their view, historical veracity appears all but irrelevant for a text and this text in particular to be normative. This narrative's normative value thus rests upon a reader's ascription to it some level of theological or, oddly enough, legal, authority. We now turn to see how these reading strategies select and interpret from narratives that are themselves selective and interpretive.

Chapter 4

Living to Die: Theo-Political Interpretations

The tyrant dies and his rule is over; the martyr dies and his rule begins.[1]

Introduction

The story of R. Chananya ben Teradyon's fiery death is relatively well known. Certainly its popularity derives from the bioethics literature, but we will discuss that in the next chapter. Here we focus on its significant theo-political familiarity. This is because it is recalled every year during the most solemn of holy days in the Jewish calendar. Inserted into the Yom Kippur "Martyrology" service is a liturgical *piyut* (poem), *Eleh Ezkarah* (These I Remember), originally composed in medieval times.[2] This poem is a list of ten sages brutally killed by Rome, among them Chananya. It includes details echoing those found in the four versions of Chananya's demise, such as a remonstrating daughter (albeit here she is a Roman legion's daughter pleading for leniency toward the victims), the rhetorical question "is this the reward for Torah?!" (albeit now spoken by *seraphim* [burning angelic guards] and not Chananya's daughter), and those wet tufts of wool affixed to Chananya's chest that are meant "to restrain himself" (*l'akev 'atzmo*). A version of this prayer, *Arzey Halevanon* (The Land of Lebanon) is recited on Tish B'Av, the holiday upon which, legend has it, many tragic events occurred to Jews, such as the decimation of both Temples, the decimation of Jerusalem, the failure of the Bar Kokhba revolt against Rome, and even more recent events like the expulsion from England, and the cleansing of the Warsaw Ghetto.

In this prayer, Chananya is listed among his colleagues as an exemplar of martyrdom.[3]

SUICIDE VERSUS MARTYRDOM

Martyrdom, and not suicide. In the case of suicide, a person intends his or her own death.[4] This cannot be assumed in martyrdom. It should be noted that the Talmud contains no particular word for suicide, and the technical discussion of purposely destroying oneself appears first in the post–Talmudic Minor Tractates.[5] This relative dearth of legal discourse does not mean, however, that characters in earlier Jewish sources did not seek their own demises. Take, for example, King Saul who was mortally wounded when fighting the Philistines. Though his guardsman refused to finish him off when so asked, Saul—fearing additional pain at the hands of his enemies—took his own sword and fell on it and his guardsman followed suit (I Samuel 31:1–5).[6] Samson also felt his blind and bound condition too burdensome to continue living, so he used his remaining strength to kill himself as well as 3,000 more Philistines (Judges 16:21–30).[7] Besieged for his treasonous usurpation of the throne, King Zimri burnt the royal palace over himself (I Kings 16:18). Several other biblical characters plead and pray for their own deaths: Moses (Exodus 32:32; Numbers 11:15), Abimelech (Judges 9:54), Elijah (I Kings 19:4; cp. Job 6:8–9), Jonah (Jonah 4:3, 8)—significant prophets and kings, no less. (Only Abimelech died because of his request; all the others died of old age or like Elijah never died at all.) Perhaps following Saul's actions, 400 Jewish children taken in a boat by Romans for probable sexual violation anticipated their own pain and reasoned that it would be better to die beforehand than suffer the intended ignominy—so they jumped into the sea and drowned themselves.[8] Just before the First Temple was destroyed, hoards of priests jumped to their deaths off the Temple's roof.[9] It is unclear whether their motivation was to escape the anticipated physical pain at the hands of the marauding Babylonians, or their stated lament for perceived lack of merit to hold the keys to the Temple—or both. Elsewhere, even social isolation was considered to be a sufficient warrant to desire one's own death.[10]

If suicide—the intentional taking of one's own life (or the request thereof)—occurs in early Jewish sources, it would be wrong to claim that the Judaic textual tradition does not countenance the behavior. A relevant ethical question is how those ancient texts and their interpretations through the centuries evaluated such behaviors. Scholars studying Jewish attitudes toward suicide uniformly agree that though suicide is morally condemnable it does not merit punishment per se.[11] They disagree on how to treat the body of someone who committed suicide, however. Some rule that anyone who knowingly commits suicide forfeits usual funereal rites (though exceptions are sometimes extended for minors and the mentally ill), while others labor to exonerate such victims and treat them with the honor due any dead person.[12] The usual key phrase in regard to one who completes a suicide is *hame'abed et 'atzmo*—one who loses oneself—an apt phrase with psychological and existential meaning.

It was the early nineteenth century sage Moshe (Chatam) Sofer who linked Chananya to the argument against suicide. To justify prohibiting suicide, Chatam Sofer focused on Chananya's *mutav* statement, the principle he taught to his students. But instead of quoting the Talmudic text, Chatam Sofer altered Chananya's words. Whereas the Talmud has Chananya saying, "It is better that the one who takes it is the one who gave it, one should not injure oneself," Chatam's Chananya says, "It is better that the one who takes it is the one who gave it; do not assume authority over oneself."[13] David Novak, the contemporary scholar who has perhaps written the most on Judaic perspectives of suicide and martyrdom, apparently concurs with Chatam Sofer's reconfiguration of Chananya's principle: "The point in this interpretation of the story . . . is that someone who has political power or authority over some other human or humans has no right to destroy or command others to destroy that human life, even if that 'other' is the victim himself or herself."[14] Whereas the Talmud's Chananya speaks of self-injury, Chatam's Chananya preaches on rightful obedience. In a Foucaultian way, deference to rightful authority matters more than—and actually, it erases and supersedes—care of the self.

That said, we should note that Chatam Sofer's version of Chananya's principle of not taking authority over oneself (*v'al yishlot hu b'atzmo*) says nothing about suicide per se. It is not, "do not lose oneself" (*v'al ibed et 'atzmo*)—the language found in the halakhic literature on suicide. Much could be speculated on why Chatam Sofer declined to incorporate that technical language here. Be that as it may, what is readily apparent here regarding intentionally taking one's life is the deliberate alteration of the classic sources to better fit and reinforce an author's position. That is, for Chatam Sofer it is acceptable to change a source so that it comports to or supports the conclusion he desires. Sid Leiman, by contrast, offers a more modern reading without altering the narrative's text. In his view, Chananya and the rabbis writ large "did not consider physical suffering a sufficient justification for suicide."[15] It is not that suffering, for Leiman, is altogether devoid of meaning, but rather it is not a sufficiently meaningful reason to seek one's mortal end.

Akin to suicide, martyrdom also entails the loss of one's life. There is a dimension of externality here, though, that is not prevalent—if at all—in cases of suicide. Whereas suicide is often private insofar as it is determined by the internal affairs of an individual, martyrdom is more public because a situation is forced upon an individual beyond his or her control. According to Novak, such dire circumstances are initiated by none other than God: "It is God who creates a situation where His glory and continued human life are mutually exclusive . . . . [I]t is not that the subject himself *feels* that his or her life is inconsistent with God's glory, but, rather, that the decision is forced upon the subject by factors outside the control of his conscious or unconscious mind."[16] Applying this theological dimension to Chananya would mean that the Roman context in which he lived—if not the Roman rules against teaching Torah publicly as well—was all divinely ordained. It would mean that Rome's regime and Chananya's demise therein were part and parcel of God's will for the world. And it would reinforce the phrase now common to describe martyrdom—***kiddush hashem***—sanctifying [God's] name. This term, however, emerged only after the third century CE, long after Chananya lived and died. Before this time there were a variety of rabbinic phrases linked to such acts.[17]

That circumstances were beyond Chananya's control has been well appreciated in contemporary scholarship. In his 1938 seminal piece on martyrdom, Louis Finkelstein described that time of Trajan and Hadrian (first–second century CE) as a "period of cultural senescence."[18] Trajan offered to rebuild the Temple in Jerusalem but quickly rescinded it for reasons too complex to unpack here. Unsurprisingly, this spurred the Jews to riot, and the brothers Julianus and Pappus came down from Syria to lead the rebellion.[19] Trajan quashed the rebellion and had Julianus and Pappus tried and eventually beheaded. Before they died, however, they retorted to Trajan, "If you will not kill us, there are many ways God can: many bears and lions are in God's world to attack and kills us. The sole reason God handed us over to you is that in the future [God] will exact punishment from you for our blood."[20] This statement is curiously similar to that said by Chananya in BT *Kallah* (see Chapter 3)—suggesting that the author(s) of these texts wanted to ensure their association, not only of context (an oppressive tyrant) but also of acceptance (assurance that their deaths are merited) as well as of theo-political response (faith in divine retribution toward the tyrant).

Their deaths further incensed the Jews of that time. Other sages quickly stood up to lead the struggle against Roman authority. After the Romans caught these new leaders and gruesomely killed them as well, they instituted further restrictions on Jewish celebrations that had "the slightest nationalist tinge," including circumcision, eating *matzah* at Passover, reading Esther on Purim, reciting the *Shema* in morning prayers, and, of course, teaching Torah publicly.[21] Lest Jews generally risked being killed on a national scale for the performance of such basic and essential Jewish rituals, the sages ruled that only three activities warranted being killed for: being forced to engage in idolatry, sexual licentiousness, and murder.[22] It was in this context of extraordinary repression designed to eradicate Judaism and Jews altogether that Chananya's encounter with the Roman government occurred.

Such a context, however, makes it difficult to clearly delineate the boundary between a death that is suicide and a death that is an instance of martyrdom. Indeed, some contemporary scholars query whether it is possible to discern an unambiguous and definitive line between the two behaviors at all, especially since their results are the same: someone is dead. Some scholars consider martyrdom to be nothing more than suicide "for a higher purpose."[23] This, obviously, leads to definitional problems as well as political ones: for who decides what constitutes higher purposes and what lower, and why should that person's (or group's) decision be the standard? Other scholars more or less acquiesce to the difficulties of discriminating between the two actions and instead conflate them. Fred Rosner, one of the premier Orthodox Jewish bioethicists, at one point claims that martyrdom "includes both the ending of one's own life for the sanctification of the name of God and allowing oneself to be killed in times of religious persecution rather than transgress biblical commandments."[24] If this definition held, self-slaughter would be treated no differently than being slaughtered by a regime—for it would be all but impossible to ascertain with any certainty precisely why a now-dead individual thought death was preferable to life. Yet later on Rosner backs away from this conflation and offers another confusing effort to distinguish them: "martyrdom, a form of suicide, is condoned under certain conditions. However, the martyr seeks not to end his life primarily but to accomplish a goal, death being an undesired side product. Thus, martyrdom and suicide do not seem comparable."[25] Given the problems such blurry definitions generate, clearly defining martyrdom becomes necessary. Martyrdom in this study means being killed for (religiously held) convictions. Though it does not entail self-initiated lethal action, an instance of martyrdom may have been sparked by personally chosen behavior known to be dangerous if not lethally criminal. That said, however, martyrdom was hardly encouraged as such within the mainstream Judaic tradition; it was "considered inculpable due to various exonerating circumstances, but it was never an approved act."[26]

It is similarly important to distinguish martyrdom from another form of killing: *mitah yafah*. This is a term associated with capital punishment meted out by a rabbinic court to a criminal deserving death. It is often justified by the biblical command, "Love your neighbor as yourself" (Leviticus 19:18)—which the rabbis interpreted to mean that an execution should be carried out quickly.[27] Though some instances of martyrdom may have occurred quickly, this cannot be assumed for all cases. Indeed, as was evident with the story of Chananya's fiery end, perhaps even a great deal of time lapses before a victim ultimately expires.[28]

This fact may have led some modern scholars to see in Chananya's demise evidence of torture.[29] Such claims are only halfway accurate, however. Torture, by definition, is meant to keep a captive alive. And torture is performed for a whole host of reasons, including securing a captive's confession for a prior crime, eliciting an epiphany or a conversion, deterring other would-be or extant reprobates, collecting corroborating material for information gleaned from other sources, and imposing upon the captive's body the complicity and docility the regime seeks among its polity in general. There are, of course, circumstances in which a captive is twisted for pain's sake alone. A captive's death is neither sought nor desired by this kind of sadist torture and certainly not by the other kinds of purposive torture. Indeed, torturers often labor mightily to keep their suffering victims alive. The Romans, by contrast, did not mean for Chananya to live; they sought his death. To be sure, their method of killing him was gruesome, as were the ways they killed other sages of that time. Thus, instead of torture, it would be better to understand that Chananya suffered a torturous execution.[30]

Theo-Political Interpretations

That Chananya died at the hands of Romans is uncontested. Precisely why he died is up for grabs, however. Many contemporary scholars portray Chananya's end as an act of theological resistance to Rome's political repression. In their view, Chananya met his fiery end not because he desired it (hence, it is not suicide) but because it was a means to achieve an ulterior purpose (hence, martyrdom). But not all scholars read the narrative of Chananya's death the same way: some emphasize the political context of the fiery scene, others examine its theological themes, and still others point to gaps or interruptions in the narratives and query their significance. We now turn to trace some of these hermeneutics, these reading strategies, of this story.

Some scholars view Chananya as an exemplar of Jewish resistance to Rome. In his essay on "Rabbis, Romans, and Martyrdom—Three Views," Gerald Blidstein examines the pedagogic purpose of juxtaposing Chananya's narrative with two other sages who idiosyncratically confronted Roman tyranny. With the printed *Avodah Zarah* edition as his textual source, Blidstein identifies two distinct narratives composed of three characters, with Chananya common to each.[31]

In the first story, R. Eleazar ben Perata and Chananya ben Teradyon were arrested simultaneously on five charges and one charge, respectively. Chananya laments to his friend that he suspects he will not survive the following legal ordeal because, unlike his friend, he only studied Torah and did not also engage in acts of loving-kindness. The narrative then moves to the courtroom into which Eleazar is brought. To each charge brought against him Eleazar responded with outright denial, obfuscation, or—miraculously—with divine intervention. In Blidstein's view, Eleazar's "activist stance is embodied by his response to the Roman tribunal: he will fight, above all, to stay alive even if he must deny that he is a loyal Jew."[32] Eleazar negotiated his way out of trouble by engaging the Roman authority on its own terms. He had no use, Blidstein asserts, "for mere demonstration, even for *kiddush hashem*"—for martyrdom.[33] Through crafty maneuvering Eleazar survived.

Chananya, by contrast, responded to his singular charge of engaging in the study of Torah with the declaration, "Thus Adonai God commanded me." The Romans immediately sentenced him to be burned, his wife slain, and his daughter consigned to a brothel. The Talmud then explains precisely why Chananya warranted death: he pronounced God's name in its full spelling, and doing so in public no less.[34] Upon emerging from the court, he declared the divine rightness of the judgment against him, citing, "*The Rock! [God's] deeds are perfect, all [God's] ways are just*" (Deuteronomy 32:4a).[35] Because Chananya does nothing to save himself from Roman execution, Blidstein sees in his response only evasion. It is as if Chananya eschews the political games Eleazar embraces and rather welcomes the political repercussions of his theological convictions. It is not that Chananya was "a man in quest of a martyr's fate," but he nonetheless ensured that his death came about so as to fulfill his own fidelity to God.[36]

The second story pits Chananya against R. Jose ben Kisma. Even from his death bed Jose has the wherewithal to confront his friend for his politically dangerous activities. Jose reminds Chananya that teaching Torah publicly is a crime in Rome's eyes, and that despite all Rome's oppressiveness it nonetheless has God's imprimatur. Chananya curtly defends himself, saying, "Heaven will show mercy." Jose then predicts that Chananya will be burned along with the Torah scroll. Blidstein sees in Jose an attitude favoring prudence: it would be better for Chananya—indeed all Jews—to toe the line of Roman rule and not overstep whatever licenses Rome extends to Jews. Chananya seemingly rejects Jose's realpolitik viewpoint that the God of revelation is also the God of history; for him, God is solely that of revelation. The Talmud seems to agree with Jose insofar as it records that it was for him and not for Chananya that even the Romans came out in droves to mourn him. On the other hand, the World to Come was Chananya's reward, not Jose's. One received earthly recognition, the other heavenly reward.

These two stories thus depict three models, or paradigms, of rabbinic activist responses to Roman rule. There was the savvy politician, the pious theist, and the acquiescent realist. One survives, one expires in flames, and one dies a natural death. Blidstein concludes that the Talmudic narrator "does not judge his characters or their political postures; each, we are led to believe, has its virtues."[37] As such, Chananya's model, like the others, stands for the contemporary audience not just as an instance of rabbinic resistance to Roman rule, but it also serves as a virtuous, admirable, even emulous exemplar.

That Chananya is to be praised for exemplary service to God and the Jewish people is reinforced by the fact that he is ensconced in the *Eleh Ezkarah* prayer recited annually on Yom Kippur.[38] Indeed, he is to be adulated for championing divine justice, for accepting his sufferings, and for acquiring the afterlife. In his analysis of the four sources of Chananya's demise (*Avodah Zarah, Semachot, Kallah,* and *Sifre*), Herbert Basser does not contest the appropriateness of seeing Chananya as a martyr. Yet, given the diverse range of stories about Chananya's fiery end and their apparent contradictions, Basser wonders whether veracity can be ascribed to any pieces of them. Can, for example, the notion that Chananya was burned with a Torah scroll be taken seriously? I will not recapitulate Basser's analysis in full (he concludes that the *Sifre* version is the earliest and original version of the story, and that a Torah scroll probably did not burn with Chananya), but rather point to Basser's assertion that "all the versions of the story invariably contain references to divine justice, or reasons for his suffering (or punishment), or an indication of his place in the afterlife."[39] That is, these stories express "theodical theologies"—theologies that wrestle with the prevalence of human (i.e., Jewish, and in particular, rabbinic) suffering at the hands of oppressive regimes. Both the prayer *Eleh Ezkarah* and Basser reconstruct history—as it is supposedly recorded in the textual tradition—so as to portray Chananya as a martyr dying for theological convictions.

Such an approach, however, pits historicity less against theology than normativity. If Basser is correct in saying that the *Sifre* is indeed not just the oldest version of Chananya's demise but the origin of the others as well, does this mean that what *Sifre* tells is historically true? If this were the case, then it is unclear whether Chananya was actually burned at all, irrespective of the presence of a Torah scroll, for that final flaming detail is absent in the *Sifre*. Insofar as the *Sifre* emerged as a codified text no earlier than the early third century CE, its written form came around 100 years after Chananya's pyre, if it happened at all. Just because a text is old does not mean it—or any other version of the story—is true.

Furthermore, given the multiple discrepancies between the *Sifre* and the other versions of the story (take, but for one example, the substitution of the executioner in lieu of the philosopher in both the *Avodah Zarah* and *Kallah* versions: each of those executioners appears with distinct moral sensibilities, suggesting they are different people altogether), questions arise whether it is feasible, legitimate, or even desirable to consider one classic text and not another as authoritative. If *no one* text can be understood as a definitively accurate portrayal of what actually happened, how can it be justified to ascribe normative authority to *any* text? Holding one text more authoritative than others would be a nearly arbitrary exercise. It also would give normative sway to a text that perhaps is more fictitious than factual.

Rebuttals to this line of critique are legion, of course. A prominent defense of viewing one set of texts as more normative than others is that historicity is all but irrelevant for norms; what matters is the social perception and reception of texts. A text's stature as normative derives not from its facts but from the factions that turn to it for guidance and structure and pass it along to future generations. Normativity is no doubt a social construct. What I suggest, however, is that there may be situations when a particular text's usual normative stature can rightfully be held suspect. The case of Chananya's demise is a prime example, for it is all but impossible to ascertain with certainty what precisely happened to him, who was around him, what he said, and who all actually died when he did. Such ambiguities discourage assuming that any one version, even the oldest, bespeaks historical truth. Rather, these ambiguities encourage a reading strategy that emphasizes theology over history, theodicy over normativity.

This line of argumentation receives support when attention is given to a glaring absence in each and every version of Chananya's fiery end. In her excellent comparative analysis of death penalty discourse among early rabbis and Christians, Beth Berkowitz notices that in this and other Jewish martyrdom narratives the Roman judge does not exist. To be sure, Roman judgment is promulgated against Chananya in each story, but nowhere does any judge's mouth utter it. Understanding these stories as complaints only or primarily about Roman authority and judgment, just or otherwise, is thus difficult to support. Rather, they can well be understood to be meditations on God's justice and rabbinic justice.

Looking closely at the *Sifre* story, Berkowitz observes that the narrative, like its neighbors in the *piska,* is "preoccupied with problems of God's justice."[40] Usually in such texts sin and punishment are correlated. Not here. The story "affirms God's justice precisely in the face of such absence [of correlation]. Moreover, punishment becomes something potentially positive, in that it provides the occasion for pious expressions of faith, and . . . it delivers Rabbi Haninah and his family in to the world to come. But we should note that it is not the punishment that earns them their merit, but their faith."[41] Consistent with the whole *piska,* Chananya and his family invoke scripture extolling God's justice, scripture that R. Judah HaNasi claims cannot be bested for this task. Wherever the divine judge may be in this scenario, divine justice pervades.

On the other hand, Chananya's *crematio*—his burning alive—recorded in the *Avodah Zarah* text may perhaps be a polemic against Rome. Two details suggest this: The Romans surrounded him with sticks, and applied wet tufts of wool to his chest. The Tannaim ruled that proper execution by burning[42] required situating a person in a pile of manure (which would serve as fuel), tie a rope around the individual's neck to force the mouth open, and then throw a lit wick into the open mouth so as to burn the person from the inside. Death would first come from within.[43] The Romans, perhaps unsurprisingly, did what they did to ensure Chananya would die from without. What the Romans did to Chananya appears "identical to that which the

Mishnah rejects," thereby proving for Berkowitz that the penalty of burning was a "battleground for the culture wars of the Rabbis."[44] Such deadly details, however, are found only in the *Avodah Zarah* text; all the other versions do not dwell on them, suggesting that different concerns are more pertinent to those sources.

It therefore seems reasonable to see in Chananya's dying stories a narrative less about politics, about rebutting Roman authority in general or questioning their execution methods in particular, than one about theology and theodicy in particular. This claim can be corroborated by the Talmudic story of another rabbi also martyred by the Romans that displays striking similarity to the Chananya chronicles.[45]

1. They said that only a few days had passed when they arrested R. Akiva and held him in prison.
2. And they arrested Pappus ben Yehudah and held him in [Akiva's] cell.
3. [Akiva] said to him, "What brought you here?"[46] He replied, "Fortunate are you, Akiva, for you were arrested on account of words of Torah. Woe to Pappus, who was arrested on account of meaningless words."
4. When they brought [forth] R. Akiva to kill him, it was the time of the recitation of the *Shem'a* [the creedal prayer]. And as they were raking his flesh with iron combs,[47] he was accepting upon himself the yoke of Heaven.[48]
5. His students said, "Our teacher, even now?" He said to them, "All the days of my life I was troubled[49] by this verse—*with all your soul* (Deuteronomy 6:5)—[which means] even if he takes[50] your soul. When will [the opportunity] come to my hands that I may uphold it? And now that it has come to my hands, should I not uphold it?"
6. He was drawing out [the word] *echad* until his soul departed with *echad* [one].[51]
7. A *Bat Kol* came and said, "Fortunate are you, R. Akiva, that your soul went out with *echad*."[52]
8. The ministering angels said before God, "This is Torah and this its reward?! *O to be among those who die by your hand, God, who die from the world*" (Psalms 17:14). God replied to them, "*Their portion is [eternal] life*" (Psalms 17:14).[53]
9. A *Bat Kol* came and said, "Fortunate are you, R. Akiva, that you are assigned to life in the World to Come."[54]

The similarities between this martyrdom story and those relating to Chananya are undisputable.[55] The absence of a human (Roman) judge, the piety of the victim, the lethal punishment for being engaged with Torah, a colleague who puts into perspective the protagonist's circumstance, a gruesome method of being killed, a discourse with present students, a statement of personal[56] theology that involves the taking of one's life, a *Bat Kol* announcing the assignment to life in the World to Come—these all suggest a type-scene, or genre, with which the ancient authors were familiar and considered a compelling vehicle to convey their martyrological theories and theodicies.

OUTSIDERS AND OUTSIDE LIFE

The theological dimension of the Chananya chronicles does not go unnoticed by the heathens mentioned in the texts. Time and again, non-Jews are visibly present and engaged with Chananya while he dies, and some even afterward. All of these outsiders are impressed with him, and all seek to associate themselves with him and his theology. The executioner in *Avodah Zarah* implores his assistance to gain access to the World to Come; the philosopher in *Sifre* would be proud to join Chananya there, too; and the executioner in *Kallah* would substitute himself for Chananya but cannot and so embarks upon what can rightly be called a fiery conversion—and he too receives the same other-worldly reward as Chananya. The only text that does not explicitly[57] present an outsider, *Semachot,* is also silent about anyone acquiring a place in the World to Come. More to the point, no other Jew in these stories—neither his family members nor his students—receives that reward. Only outsiders accompany Chananya outside life.

Why would outsiders be attracted to Chananya? Is it for his charisma, his piety? Or perhaps it is because he is willing to undergo martyrdom by a tyrant's decree in the first place. Søren Kierkegaard, himself a sufferer for his faith in nineteenth-century Denmark, notices that though they are enormously different, both a tyrant and a martyr share "the power to compel." He goes on:

> The tyrant, himself ambitious to dominate, compels people through his power; the martyr, himself unconditionally obedient to God, compels others through his suffering. The tyrant dies and his rule is over; the martyr dies and his rule begins. The tyrant was egoistically the individual who inhumanly made the others into "the masses," and ruled over the masses; the martyr is the suffering individual who educates others through his Christian love of mankind, translating the masses into individuals—and there is joy in heaven over every individual whom he thus saves out of the masses.[58]

Martyrdom individuates; tyranny generalizes. It could very well be that these outsiders who are willing to die so as to live in the World to Come desired this outcome precisely because they could not fathom existing any more in a regime that denied them their names. Note that none of them in these texts has a name, only a role.[59] Indeed, namelessness could be commentary against Roman governance expressed by the authors of these martyrological stories insofar as the Jewish victims are named. Still, these particular Romans somehow felt that they could achieve and perhaps even enjoy their individuality only beyond generic Roman life in this world.

The notion that gentiles can have a place in the World to Come is well ensconced in the Judaic textual tradition.[60] And the notion of Roman legionnaires and leaders interested in gaining access to life in the World to Come is also well founded, as evidenced in this Talmudic story.[61]

1. When Turnus Rufus[62] the wicked destroyed[63] the Temple,[64] Rabban Gamliel was decreed to die.
2. An officer[65] came and stood in[66] the House of Study and said, "The Nose-man[67] is requested. The Nose-man is requested."
3. Rabban Gamliel heard and went and hid himself.
4. Thereupon the man went to him in secret and said to him, "If I save you, will you bring me to the World to Come?" He replied, "Yes." He said, "Swear it to me." He swore to him.
5. [The man] ascended to the roof and fell and died.
6. There was a tradition[68] that when a decree is made and one of them died, then the decree against him is annulled.
7. A *Bat Kol* came and said, "This[69] officer[70] is assigned[71] to life in the World to Come."

It is improbable this incident actually occurred, since Rabban Gamliel predated both Turnus Rufus's leadership and the destruction of the Second Temple by perhaps as much as a hundred years. Regardless, it seems the story is more concerned about the fate of Turnus Rufus than Gamliel's. Though Sextus Julius Severus finished suppressing the Bar Kochba revolt in 135 CE, it was Turnus Rufus who began that Roman operation a few years earlier, and apparently his conscience was troubled by his own overpowering onslaught. Knowing that Roman magisterial edicts were in force only during the lifetime of the promulgating magistrate (paragraph 6 above), Turnus Rufus figured that his own rulings could die with him. Thus, perhaps in a disguise as "the [anonymous] man" (4), Turnus Rufus secretly offered Gamliel a reprieve: he would exchange his own life so as to protect Gamliel's, as long as he was guaranteed by Gamliel access to the World to Come. The emergence of the *Bat Kol* expressed divine approval of this plan.[72]

Whether gentiles sought access to the World to Come because of problematic personal identities within the Roman empire (e.g., Chananya's executioners), or because they had troubled consciences due to their own excessive military repression (e.g., Gamliel's Turnus Rufus), such stories bespeak a kind of theological triumphalism. Rome may have had the upper political hand in this world, but it was the Jews who served as gatekeepers to theological rewards beyond this world. Such stories depict Romans seeking, begging for, and even coveting Jewish theological approval. Historicity is of little concern, it seems, to the authors of these stories. Rather, by painting Roman authority figures in these desperate hues such stories provide a morale boost for fellow Jews suffering at the hand of an oppressive foreign regime.

DIVINE IMPRIMATUR?

That said, such links between outsiders and the World to Come in the Chananya chronicles warrant another look. Why, for example, does a *Bat Kol*—a heavenly voice—interrupt the otherwise human flow of the story? What is the meaning of this divine declaration: is it a commendation of what the characters have done insofar as it announces a coveted reward bequeathed upon them? Or might it be something else?

Recall that the reason Chananya was to be killed is patently unclear. Whereas *Semachot* insists he was guilty of heresy, *Kallah* points to his admixture of charity monies, and *Avodah Zarah* adduces to him the crimes of teaching Torah, speaking the name of God aloud in public, and mixing charity monies, *Sifre* by contrast says nothing about why he merits death. Note that the World to Come appears in all except *Semachot*, suggesting perhaps that no heavenly reward accrues to those engaged in heresy. And why might heresy be distinct from other crimes in this regard? Unlike those other activities, heresy perverts religious doctrine, and it does not necessarily involve other people. Perhaps heaven's silence reflects divine disparagement for such untoward theological activity, especially by someone who purportedly is pious. Conversely, this otherworldly reward is mentioned precisely in those situations where Chananya's actions necessarily affect other people: charity funds, teaching Torah, and speaking God's name publicly.

To boot, these more public actions reflect and reinforce mainstream Judaic religious practices. Insofar as the rabbis of the Talmud in general sought to clarify piety and demonstrate its dimensions, it would only be logical to tack on to certain stories involving these public gestures of proper piety some indication of divine commendation for them.

The source of this divine approval remains suspect, however. Its arrival comes either through the mouth of a gentile, such as the philosopher in *Sifre,* or through the mouth of the *Bat Kol,* as in *Avodah Zarah* and *Kallah.* How much credence can be given to these sources? In regard to the philosopher, his is an expression of gentile theological theodicy and political pessimism. Insofar as he rebuked the governor for executing Chananya, it would seem that both Roman governance and its attendant paganism no longer sufficed for him. He would rather throw in his lot with the condemned because Jewish theology offered him greater security than the alternatives. Nevertheless, given that this philosopher is so willing to abandon his own theo-political paradigm, could he be considered a credible source regarding the purported reward showered upon Chananya? One might argue that he experienced some kind of conversion to Judaism and thus his testimony deserves consideration. This position lacks evidence and therefore can be disregarded. So on the other hand, one might contend that though the philosopher may not have fully converted, he at least intellectually understood and appreciated Chananya's piety and plight; his statement about the World to Come would thus serve as a belated barb at the ridiculousness of executing this individual—and people generally—whose resistance to Rome transcends its hungry imperial flames. The philosopher sees vacuity in Roman rule and admires the substance of Jewish theo-political resistance. Whatever his new-found allegiance to the underdog might be, it does not automatically render his statements about Jewish theological rewards true.

As for the *Bat Kol,* that voice, too, is sketchy. Despite its pervasive presence in rabbinic literature, its authority in human affairs remains doubtful. Take, for example, the famous narrative of the Oven of Akhnai.[73] This story involves a heated dispute between R. Eliezer and his colleagues about the ritual purity of the oven, the details of which are well rehearsed and analyzed.[74] In brief, Eliezer performs miracles to demonstrate the rightness of his position. Time and again his colleagues refuse to accede to his perspective and hold their own. He finally resorts to invoking the ultimate proof: heavenly support for his position. A *Bat Kol* appears and declares that the law is as Eliezer declares it. Immediately R. Yehoshua stands up and quotes, "*It is not in Heaven*" (Deuteronomy 30:12), which R. Jeremiah interprets to mean, "Since the Torah was given at Mt. Sinai, there is no need to heed a *Bat Kol*, for it has been written, '*follow the majority*' " (Exodus 23:2). While many scholars who quote this narrative end here, the story continues. Some time later Elijah met with R. Natan, who inquired of the prophet what God did at that critical moment when R. Yehoshua stood up with his proof-text rebuttal. Elijah replied, "God smiled, saying, 'My children have defeated me, my children have defeated me." Without getting too sidetracked into the logical depths of this fascinating story, suffice it to say here that a *Bat Kol*'s declaration does not necessarily mean that whatever is said is so, or even that it *should* be so.[75] If the *Bat Kol* cannot be trusted in *every* instance as an indicator of what is or what ought to be, it cannot be relied upon without suspicion in *any* scenario. Thus it is only logical that what the *Bat Kol* says in the Chananya chronicles should be held suspect.

Normal or Exceptional Death?

If the statements of the philosopher and the *Bat Kol* cannot be accepted without some serious reservations, what normative conclusions can be derived from these readings of the Chananya chronicles? Are *any* normative conclusions possible? Answering these questions requires invoking another Talmudic dictum relevant to this discussion. The sages rule that, "No halakhic matter may be quoted in the name of one who surrenders himself to death for the words of Torah."[76] This principle is critically relevant to the case of Chananya. According to Abraham Gross, "A survey of the Talmudic literatures shows that not a single saying in the name of R. Hanina ben Teradion survives in the tradition (with the possible exception of BT *Menahot* 54a). Could this lacuna reflect an attempt on the part of some of the Sages to uproot R. Hanina's norm of martyrological conduct?"[77] Or it could be read the other way: since no norm is recited in Chananya's name, he is a reasonable character to deploy in narratives that wrestle with the theo-politics of martyrdom and suicide. Either way, the consistency between this principle and the lack of Chananya-based rulings in the Talmudic corpus suggests that perhaps it would be wrong for post-Talmudic norm creators to lean heavily on Chananya—be it here in regard to death and dying, or on any other issue, for that matter.

This consistency between the principle and lack of any ruling by Chananya is put into stark relief when we recall the ambiguities and inconsistencies of the various stories surrounding Chananya's demise. Not least in regard to the *Avodah Zarah* text in which myriad complexities and inconsistencies exist, Jan van Henten "wonders whether R. Hanina functions as a model figure at all."[78] Indeed, that text incorporates a conundrum that defies extracting a clear and definitive norm. "If one can obtain [eternal] life by a deliberate act of self-destruction [as did the executioner], what happens to Hanina's original statement that one must not even hasten death, much less directly destroy oneself?"[79] If Chananya himself cannot serve as a model, how reasonable is it to consider his evidently abrogable teaching as a norm?

Such principled, textual, theological conundrums make it difficult to point to Chananya as an exemplar of martyrological fervor, though some scholars do.[80] Sidney Goldstein argues that Chananya

> perpetrated an act of intense piety (*midat chasidut*) as he understood it and as his individual needs required . . . [He prolonged his own suffering, which was] then a *personal* act of piety, not necessarily an act governed by specific Judaic precepts. This act of intense piety (*midat chasidut*) is a unique and highly personal phenomenon. It extends beyond religious requirements. It is difficult to base a halakhic decision only on a precedent which is basically an act of intense personal piety.[81]

If Chananya's demise was a personal expression of extraordinary piety, his can rightfully be understood to be exceptional behavior. Such supererogation is called in Hebrew *lifnim meshurt hadin*.[82] The exceptional, however, is a category that transcends sociality, the realm in which norms operate. As Kierkegaard observes, "I cannot possibly make it clear enough that the Exceptional has nothing whatsoever to do with ethics; ethically there is nothing exceptional, for the highest is quite simply what is demanded . . . . The 'Exceptional' has nothing to do with ethically fulfilling what is demanded, but is a particular relation to God."[83] Chananya's exceptionalism pertains more to his personal relationship with God; its relevance to fellow humans is all but nil.

How, then, could his actions or his words justify *un*exceptional norms for the public? Regardless of whether it could or could not, perhaps it should not. This conclusion is supported by the Talmud itself insofar as some extant *Avodah Zarah* manuscripts have Chananya insisting that what he chooses to do is for himself alone and should not be considered a norm applicable to others who might find themselves in similar circumstances.[84] And as just noted, the Talmud records no halakhic teaching in the name of Chananya ben Teradyon. For all these reasons—competing narratives, diversely motivated insiders, suspiciously motivated outsiders, dubious divine imprimatur, and apparent exceptionalism—it is therefore reasonable to conclude that pointing to Chananya, to his life, or much more to his dying and death, cannot serve a theo-political normative purpose.

Chapter 5

Dying to Die: Bioethical Interpretations

It is precisely by reason of this entanglement, as much as by being open-ended on both sides, that life histories differ from literary ones, whether the latter belong to historiography or to fiction. Can one then still speak of the narrative unity of life?[1]

Introduction

Just as deriving sociopolitical norms regarding martyrdom from the Chananya chronicles is difficult at best, so too is the task of extracting bioethical norms from those sources. Still, contemporary bioethicists invariably turn to this story when they expound upon the dilemmas surrounding euthanasia. Not only do they look to Chananya's death for guidance, but they also frame their ultimate conclusions regarding euthanasia based on how they read the story of his dying. In this way Chananya's dying and death become the exemplars par excellence, the models after which and against which subsequent end-of-life care is and should be shaped and measured. In the view of many scholars, contemporary medical care at the end of life should comport to how a Jewish bioethicist reads this story.

Such reliance upon this story inevitably ascribes to it extraordinary power. Its repeated invocation by bioethicists perforce demarcates it as a norm-generating text. For this sociological reason alone, deference to it appears reasonable. Yet, should this be so? No contemporary Jewish bioethicist writing on euthanasia today pauses long enough to consider why it is that this peculiar story was brought into the conversation in the first place. While all bioethicists point to the Talmudic *Avodah Zarah* version as the primary source for this story, few make the effort to locate it within its larger context, and hardly more than a couple acknowledge that competing versions of Chananya's death exist in other classic sources. And none explicitly justifies how and why she or he reads the story as such, ending it here and not there, highlighting this detail but not that one. A variety of questions thus merit attention regarding the presence, use, and normative stature of (bits of) this story, that is, questions of prevalence, hermeneutics, and salience.

Before we can attend to these questions, we should take a moment to recall the narrative nature of Chananya's demise. As shown in Chapter 3, every version of his death is a story. A story, or narrative, necessarily involves connecting at least two events, and through that very connection each event is interpreted against the overarching whole. Be it his reaction to the verdict promulgated against him and the reactions of his colleagues and strangers (e.g., *Sifre Devarim* §307), or the conversations he has with the executioner (e.g., BT *Avodah Zarah* 18a; BT *Kallah* 51b), or the intimate conversations he has with his daughter (e.g., BT *Semachot* 8.11)—all versions stitch together at least two events. In so doing time must pass. Stories thus depict time and change.

They also entail inaccuracies since they present, or better, re-present, those events from particular angles and perspectives. In their retelling of events, narratives highlight certain details and gloss over or skew others. This should not be surprising, for, as the primatologist and neuroscientist Michael Gazzaniga has found out, the interpretive center in a brain reconstructs events and "in so doing makes telling errors of perception, memory and judgment. The clue to how we are built is buried not just in our marvelously robust capacity for these functions, but also in the errors that are frequently made during reconstruction. Biography is fiction. Autobiography is hopelessly inventive."[2] Such cynicism, if left alone, could fester and undermine the value of this project, as well as the bioethical practice of turning to both biography and autobiography when wrestling with the ethics of complex cases or issues. Instead, we should acknowledge that this fact of inherently errant cognition does not mean that all (personal) narratives or their constituent details are completely non-factual. Nor does it mean that narratives in whole or in part have no value whatsoever when discerning truth or when constructing norms. The challenge is to justify imputing to a narrative, or to particular details of a narrative—whether biographical or autobiographical—truth or normative power.

Complicating this task is what Alasdair MacIntyre observes about all characters—whether real or fictional: they "never start literally *ab initio;* they plunge *in medias res,* the beginnings of their story already made for them by what and who has gone before."[3] So finding precisely the actual start of either a character or a story is difficult, be it in a text and especially in life. This pre-text, if you will, of a person, character, or story, also merits consideration, as much as the post-text, that is, what happens after a story or life ends. Allocating a particular moment as the start of a specific story is for the most part an arbitrary exercise, for it could be assigned to another moment and not to this one where the ink appears on the page or to when that balloon bounced around a specific corner on the busy street.

In the context of Jewish bioethics, it would be reasonable to respond to these challenges by claiming that the Judaic textual tradition is what it is: its stories begin when and where they do; there is no pretext and no arbitrariness is involved. This rejoinder fails, however, to grapple with the truth of MacIntyre's observation that all stories—indeed all lives—are entangled and mutually limiting:

> The difference between imaginary characters and real ones is not in the narrative form of what they do; it is in the degree of their authorship of that form and of their own deeds. Of course just as they do not begin where they please, they cannot go on exactly as they please either; each character is constrained by the actions of others and by the social settings presupposed in his and their actions.[4]

Whereas real people have some level of agency to direct the flow of their life's plot and literary characters do not, both share the trait that their stories are not their own, exclusively. This leads some, like Paul Ricoeur, to wonder if it is even possible to differentiate between the literary and the lived.[5] Both factual and fictional lives are by definition and by necessity caught up in larger contexts that shape, frame, and constrain them. Appreciating those contexts would, I venture, only enrich one's understanding of what happens within and through the events recorded in a particular story.

In sum, any and every story—including the story/stories of Chananya's death—incorporates time, change, and a larger context (both before and after) in which that time and those changes occur. Allocating to the story normative capacity, or what Robert Cover calls jurisgenesis, requires identifying within it that which is universal and unyielding, since norms like laws are meant to be both. This story is neither universal nor unyielding, though. It is peculiar insofar as it relates the story of a specific individual embroiled in a particular historical circumstance, and it seems he cherished certain and not other values that perhaps were not widely shared. Ascribing to the story universality is thus questionable at best. And the story yields, that is, to the degree that the main character continuously shifts his position about the question at hand—the appropriateness of intervening in his own dying process—the narrative's clear and definitive stance is elusive. This is undeniably true in regard to the *Avodah Zarach* version most commonly cited by bioethicists. One wonders whether extracting a solid norm from this wiggly and curious narrative to guide contemporary bioethical practice is possible or even desirable.

The foregoing notwithstanding, contemporary Jewish bioethicists invariably turn to BT *Avodah Zarah* 18a as the *locus classicus* for constructing norms regarding modern Jewish attitudes toward and practices of end-of-life care, especially euthanasia. But why BT *Avodah Zarah*? Why not one of the other sources or a selection of them? Even if modern scholars knew of these competing versions found in other sources (*Sifre Devarim, Kallah, Semachot*), the several reasons that they exclusively invoked BT *Avodah Zarah* are understandable. The most obvious is that the Babyalonian Talmud has long been considered throughout the Jewish world as the most authoritative classic rabbinic text; all others pale in comparison. Furthermore, the Babylonian Talmud includes within its sprawling corpus a great deal of legal discourse and legal decisions—along with history, geography, science, medicine, zoology, astronomy and astrológy, magic, genealogy, as well as innumerable parables and stories, prayers and rituals, among other genres. Certainly other collections of rabbinic texts may touch on the law, as does *Sifre Devarim* since it is a collection of *halakhic midrashim* or legal stories, or offer legal decisions, as do *Semachot* and *Kallah;* but their relative weight in normative deliberation is slight, especially when countered by the Babylonian Talmud.

This may explain why modern bioethicists look exclusively at the Babylonian Talmud's *Avodah Zarah* version of Chananya's demise, but it does not explain why they look at Chananya's demise in the first place while they are making norms regarding euthanasia. When did this practice begin? A partial answer can be found in one of the earliest—and most famous—Jewish bioethical texts published in English, Immanuel Jakobovits's 1959 *Jewish Medical Ethics.*[6] There, in the chapter on "The Dying and Their Treatment," and under the heading, "Treatment of the Dying," Jakobovits cites this *Avodah Zarah* text as the source of the "uncompromising opposition to any deliberate acceleration of the final release" of a dying patient.[7] Just a few years before this in 1950 in the United States, Israel Bettan composed a Reform *responsa* (a rabbinic answer to a perplexing moral question) on the question of euthanasia and included a reference to this story, though here it was invoked to demonstrate what the "religious man's attitude toward a life of affliction should be. He will accept the lot apportioned to him."[8] Prior to these mid-century publications, this *Avodah Zarah* story was most often referenced in regard to martyrdom and suicide.[9] Thus this story's inclusion in biomedical discourse on euthanasia is relatively new.

Details, Details

That modern Jewish bioethicists would turn to the Babylonian Talmud over other sources and would include this particular story about Chananya's fiery end in their thinking on euthanasia are understandable moves. The next task is to see how they read this peculiar story. What we will find is that while there are many details of the story that could be considered, modern bioethicists choose to hone in on only a few of them. The other details are ignored altogether, glossed over with truncated summaries, or misconstrued so as to reinforce a scholar's preferred reading of the whole. Such reading strategies are to be expected, however, and we will discuss why further on, especially in Chapter 6. The task at hand is to identify and understand how and why the selected details are read and used.

Before turning to those specific details, though, we should situate them in their context, that is, in the entire narrative. To this end here is the full story again as it occurs in the Babylonian Talmud, *Avodah Zarah* 18a (printed edition; for manuscript variants, see Chapter 3)—the version that all bioethicists invoke in their essays on euthanasia.

1. They said that only a few days had passed since the passing of R. Yosi ben Kisma when all the elite of Rome went to bury him and eulogize him with a great eulogy.
2. Upon returning they found R. Chananya ben Teradyon who was sitting and engaging in Torah, convening substantial gatherings, and a Torah scroll rested on his lap.
3. They brought him and wrapped him in the Torah scroll, and they placed piles of twigs around him, and lit them aflame. They brought tufts of wool soaked in water, and placed them upon his heart so that his soul would not depart quickly.

4. His daughter said to him, "Father, to see you thus!" He replied to her, "Were I to be burned alone it would be a difficult matter for me. Now as it is me who burns and the Torah scroll is with me, the one who will address the affront to the Torah scroll will also address the affront to me."
5. (a) His students said to him, "Teacher, what do you see?" He replied to them, "The parchment burns but the letters soar." (b) "Also you can open your mouth and the fire will enter you." "It is better that the one who takes it is the one who gave it. One should not injure oneself."
6. The executioner said to him, "Teacher, if I increase the flames and remove the tufts of wool from your heart, will you bring me to the World to Come?" He said to him, "Yes." "Swear to me." He swore to him. Thereupon he immediately increased the flames and removed the tufts of wool from his heart. His soul quickly departed. He even jumped and fell into the fire.
7. A *Bat Kol* [a heavenly voice] came and said, "R. Chananya ben Teradyon and the executioner have been assigned to life in the World to Come."
8. Rabbi [Judah HaNasi] cried and said, "Some acquire his [eternal] world in one moment, and some acquire his [eternal] world in many years."

The details Jewish bioethicists predominately fixate upon are these three: (a) Chananya's teaching to his students, which we call the *mutav* statement (paragraph 5 above); (b) the Roman executioner's words and actions (6); and (c) Chananya's response to the executioner (also found in 6). There are exceptions to this pattern, of course: some bioethicists notice other details in the story and discuss them. These more expansive readings nonetheless incorporate consideration of the three central details. Could it be said that these three details are necessary and perhaps even sufficient for a modern Jewish bioethical argument on euthanasia? Such a reductionist or essentialist claim would be dangerous, to be sure. Indeed, one goal of this project is to show that these three elements may be more distracting if not damaging when contemplating end-of-life care for a loved one; other features of this story may be more relevant. Those suggestions will follow the analysis and critique of current reading strategies of these three details.

Mutav

The first detail in the story modern bioethicists invoke—though perhaps not the first one mentioned in their arguments—is the interaction between Chananya and his students. Curiously, scholars do not contend with the totality of this interaction but only a portion thereof.[10]

His students first ask what he sees (see paragraph 5a above). This seems to be an odd question given the circumstances.[11] Their beloved teacher, bound and wrapped within their holiest of texts, is encased in flames. And their first impulse? It is not to save him or to retrieve the Torah scroll. As if oblivious to his agony—both physical and metaphysical—they seek only to extract from him one last insight, a final illuminating perspective on what it means to exist in and exit from this world. Now one might think they are being callous; but recall that they just witnessed Chananya's conversation with his daughter. To her he abjured metaphysical discomfort; he is content with his fate and encourages her to find solace in faith as does he. It thus would be reasonable for the students to take advantage of his equanimity and continue to treat him as they would in any other situation, with all the reverence due to their master teacher. Their nonintervention thus makes modest sense.

And so what does Chananya see? "The parchment burns but the letters soar." He sees the great mystery of revelation ascending back to its source while its physical manifestation returns to the very ashes his body will also soon become. God's revelation is no material thing vulnerable to destruction as is a human body. Flames cannot annihilate faith. Or perhaps he sees the language of holiness fluttering away precisely in this moment of extreme need. Its effervescence and immateriality appears impotent against the solidity and earthiness of fleshy human existence. However elevating it may be, faith cannot douse political fires. Indeed, faith and its fleeing language necessarily fail when skin (human or animal) is on the line. However inspiring this language and theology might have once appeared, its inked instantiation means nothing unless and until it is shaped on something tangible in this world.

If this is what he sees, what does he want of or from his students? It is far from clear. He could be instilling upon them the importance of holding fast to their faith despite and because of imbroglios. Resist not in the flesh but in spirit, he seems to say. Or perhaps: take on a quietist political stance in this world. Or maybe he would rather they abandon their flights of fanciful faith and raise their fists in constructive protest. Be real, be here and now, burn with rage as does this skin of mine and the skin of our cherished Torah; release your faith to the whims of the wind and physically engage the politically mighty.

While certainly there are many other ways to read this initial exchange between Chananya and his students, this very polysemy may be the primary reason that modern bioethicists avoid referencing it. Indeed, nowhere in contemporary Jewish bioethical discourse are these words quoted much less meditated upon. Rather, bioethicists avoid this perplexing exchange and hone in on the next part of the conversation.

After hearing what he sees, Chananya's students implore him with a proposal to asphyxiate himself (see paragraph 5b above). He responds, "It is better that the one who takes it is the one who gave it. One should not injure oneself" (*mutav sh'yitalnah mi sh'nitnah, v'al yechavel hu b'atzmo*). It is unanimously understood among bioethicists that the "it" here refers to human life—though some refer to "it" as the human soul. While a few scholars invoke this plainest reading of the text, many if not most do not: they alter it to better fit their preferred conclusions. And to achieve this they manipulate the phrase in a variety of ways.

One significant way bioethicists manipulate the phrase involves the initial comparative form (*mutav*, or "it is better"). Take, for example, Joseph Prouser's version, "Let God who gave life take it away; let no mortal commit self-injury."[12] Similarly, Byron Sherwin, David Goldstein, and Steven Resnicoff offer, "Let Him who gave me [my soul] take it away; no one should injure oneself."[13] Such language of "letting God" suggests—and not unambiguously—that humans have the capacity *not* to let God do something. It implies that humans, even frail and especially enflamed humans, could prevent God in some way or another. Such human potency may be desirable, especially when it comes to care at the end of life. But is it such potency that Chananya says or even implies? Is he really making a comment about the capacity of humankind to thwart God's will? Perhaps, instead, Chananya meant what Nisson Shulman insists he said, "Only He who has given life may take it away. No one may hasten his own death."[14] Here Chananya's judgmental *mutav* mutates into a declarative mood expressing an unqualified statement of fact. Whereas the actual Talmudic text bespeaks Chananya's confidence in his statement, in Shulman's version his confidence is irrelevant.

Jakobovits maneuvers the *mutav* differently. For him Chananya states, "It is better that my soul shall be taken by Him Who gave it than that I should do any harm to it on my own."[15] This configuration juxtaposes God's agency to a human's, specifically Chananya's, and discerns that divine action trumping human action is for the best—whatever that may be. This reading, however, dismisses the Hebrew *v'al*, which is rendered here as, "One should not." The preliminary *vav* serves as a disjunctive; it separates two complete thoughts into two distinct, yet related, sentences. Perhaps an even stronger translation would be, "But one should not . . . "[16] By eliminating the disjunctive, Jakobovits's *mutav* construes human agency on nearly the same level of potency as God's. Again, is this what Chananya actually meant?

A second strategy bioethicists use to alter the *mutav* phrase is through obfuscation by truncating or vaguely translating it. Yitzhok Breitowitz clearly demonstrates truncation with his, "Let He who gave me life take it."[17] And Ron Green's—"Only He Who gave life can take it away; I may not do it myself"[18]—fluctuates with vagueness: for what does the second "it" mean? Given the context of his argument, this "it" refers to suicide, which, as shown in Chapter 4, is a forced reading of the text.

Third, and even more distant from the phrase than the second way, is the offering of an interpretive gesture toward it. For example, J. David Bleich nods toward the statement by claiming, "R. Hanina explained that such an act is forbidden."[19] According to Bleich, Chananya proscribed specifically the act of opening his mouth to the fire so he would be spared further agony—as suggested by his students. Rosner also gestures toward Chananya's actual utterance in this summary: "When his disciples pleaded with him to open his mouth so that the fire would consume him more quickly, he replied that one is not to accelerate one's own death."[20] But did Chananya actually say that such an action was forbidden, or did he insist that other principles (e.g., God as giver and taker of life; one ought not injure oneself) are at play? Such gestures toward Chananya's *mutav* statement thus appear to be more violent than benign: they move the argument along as the authors desire but not as the *sugya* designed.

Another version of this impulse is Walter Jacob's view that Chananya's statement is liturgical: "Though Hanina was unwilling to pray for his own death as his disciples suggested . . . "[21] Even though praying for death has biblical roots and rabbinic branches, and Chananya himself earlier praised the justness of the punishment he is to receive (as summarized in Chapter 3), it is as difficult to construe this *mutav* statement as prayer per se as much as it is to claim that the students' proposal was one for prayer.[22]

The fourth and most prominent way bioethicists contort the *mutav* phrase is through the final verb, "to injure" (*yechavel*). There are, to be sure, some bioethicists who retain the verb as meaning self-injury. Avram Reisner offers an illustrative felicitous translation: "It is well that He who gave it should take it. One should not injure himself."[23] Most bioethicists, however, tweak *yechavel* to mean what they want it to say. Several conflate self-injury with self-destruction, such as David Novak's, "It is better that He who gave life Himself take it back and let not a man destroy himself."[24] It is unclear precisely what self-destruction means, for it could mean self-mutilation (e.g., scarring oneself), self-debilitation (e.g., amputating a limb), or self-killing (e.g., suicide). Logic suggests that Novak means lethal self-destruction, but if so, why did he not say it explicitly?

Most scholars avoid such ambiguity. They insist Chananya's *mutav* proscribes hastening one's death: nothing less and nothing else. Shimon Glick offers a typical version of this reading, "Better that the Lord who gave me my life take it from me rather than that I should contribute to my demise."[25] Sometimes bioethicists make the final words be "hasten one's [my own, his own] death."[26] Goldstein's is glib: "and one should not kill himself."[27] Such modern versions might have taken their cue from Julius Preuss's 1911 German translation, "*es ist besser, dass der, der mir das Leben gegeben, es mir auch nimmt, als dass ich mich selbst umbringe,*" that Rosner renders as, "It is better for the One who gave me life to take it away from me, rather than I should bring about (i.e., hasten) my death."[28] (Perhaps a better translation of the concluding phrase would be: "than that I should kill myself.") The question remains open whether this is what Chananya actually said.

The Talmud elsewhere has lengthy discussions about injury (*chovelut*—the same root as Chananya's *yechavel*) and assault, yet it does not mention that these activities involve lethal injury or lethal assault.[29] Self-injury, at least Talmudicly understood, is a non-lethal activity. A further question is where in Chananya's *mutav* these bioethicists see the notion of hastening, quickening, or accelerating death. Recall that his actual words—according to those recorded in the Talmud—were, "one should not injure oneself." These facts thus render suspect those bioethical arguments that construe Chananya saying here that one should not hasten one's death. Such so-called translations are more political than literal insofar as they promote a scholar's desired argument and preferred conclusion than they provide an accurate transcription of the original Talmudic source. They also prime the reader, since certain normative conclusions will appear logical—even foregone—given these "translations." For example, if it were truly the case that Chananya instructed his students that one should not hasten one's own death, a bioethicist's proscription of both active and passive euthanasia seems consistent with that teaching, and a reader of that essay would readily go along with the bioethicist on this point. Again we must ponder if and how such rhetorical strategies maintain fidelity to the textual tradition, or if they reflect other agendas.

Another strategy is found among those scholars who offer a personalized or reflexive rendition of Chananya's *mutav* (e.g., "I should not injure myself," or "I will not hasten my own death"). Curiously, none of them cites the manuscript traditions of the Babylonian Talmud in which such statements actually occur.[30] Rather, they impute to him a reflexive commitment that does not accord with the text they consult—the Vilna printed edition. Not only that, they then invariably assert that his personal commitment not to assault himself (lethally) applies to everyone as would the generic principle that he does articulate ("one should not . . . "), according to the Vilna edition. That is, these bioethicists put words into Chananya's mouth that he does not say yet simultaneously they do not follow the logic of those purported dying words.

Such a careful analysis of how the *mutav* statement is read is required because time and again, scholars point to it as the pivot around which their arguments turn. In regard to suicide, many like Reisner consider Chananya's teaching to be "the basis of the Jewish and the general prohibition on suicide, a prohibition that the ethicists are loath to lose even as they function under the rubric of perfect autonomy."[31] Yaakov Weiner points to the *mutav* statement as "a source for the ruling that no matter how much one suffers, suicide cannot be condoned or allowed."[32] Baruch Brody points to R. Tam's reference of the 400 children who jumped into the sea to prevent being abused sexually. Whereas these children had a reason to end their lives, Chananya, "was no longer being offered the choice between life and sinning. He was condemned to death, was being killed, and did not have to fear that he would apostasize to save his life. He had therefore no justification for hastening his death. The four hundred young people were different. They still had to confront the choice between sin and death. They had to fear that they might sin under torture. Their decision to commit suicide was justified."[33] In Brody's view, hastening one's death without a just cause is a prohibited form of suicide. Novak extends this to mean that physician-ordered suicide is prohibited: "Following the implication of Rabbi Hanania ben Teradyon's answer to his disciples, we can better understand why a physician has no right to order somebody to kill himself or herself and why the patient who receives such immoral orders has a duty to resist them."[34] Again we should pause to inquire whether this is what Chananya actually said, and if not, is this what he meant by what he did say. At first glance, conclusions such as Novak's appear to be forced since Chananya says nothing about ordering someone else to commit suicide.

Other scholars see in Chananya's teaching a proscription of assisted suicide. Assisted suicide is a situation wherein a care provider knowingly offers a patient the wherewithal to end his or her life, though the provider does not administer those means directly. Nisson Shulman, for example, declares that Chananya "refused to open his mouth, because suicide and euthanasia are forbidden"—and since these are forbidden, "physician-assisted suicide is causative murder."[35] Joseph Prouser similarly asserts, "Suicide is forbidden, Hanina taught us through both deed and dictum, even for one suffering pain and illness. It is readily apparent that this prohibition extends to physician-assisted suicide."[36] Eliezer Waldenberg extracts from this *mutav* statement that physician-assisted suicide is proscribed and that everything possible that can be done to prolong a life must be done.[37]

If suicide and assisted suicide are proscribed by this *mutav* statement, so would be active euthanasia, whereby a care provider intentionally intervenes in a patient's dying process in ways that brings about that person's death.[38] Jakobovits offers a robust version of this argument: "This uncompromising opposition to any deliberate acceleration of the final release is well exemplified by the martyred sage Hanina ben Tradyon who, whilst the Romans burnt him at the stake, refused to follow his disciples' advice to open his mouth to the flames (in order to speed his death) with the defiant proclamation: 'It is better that my soul shall be taken by Him Who gave it than that I should do any harm to it on my own.' "[39] As noted above, Jakobovits alters Chananya's dying words from a generalized principle others might follow into a first-person commitment—but then he reads that idiosyncratic conviction as a general rule. Why Jakobovits imposes upon the text and then does not follow its new logic is a curious if not specious move.

And finally, some scholars read in this *mutav* statement not merely a prohibition against active euthanasia, but also a similarly strong proscription of passive euthanasia. Whereas many other scholars acknowledge that from the Chananya story some permission could be found for passive euthanasia—such as removing impediments to dying—Weiner asserts that these are "misinterpretations" of the texts.[40] In his view and following a reading of Maimonides (MT *Hilchot Rotze'ach* 2:2, 3; 3:9), any effort that either hastens a person's death or does not prolong a person's life is murder. Thus, given Chananya's statement, all kinds of euthanasia are prohibited—active as well as passive.

While the impulse is strong among many contemporary Jewish bioethicists to construe Chananya's words to fit agendas other than what is found in the Talmudic text, some scholars resist this temptation and offer fairly accurate translations. Take, for example, Elliot Dorff's version: "Better that God who gave life should take it; a person may not injure him/herself."[41] Though Joseph Telushkin's—"It is better that He [God] who gave my soul should take it, and let no person harm himself"—also stays close to the original text, he interprets it to mean that Chananya was "at first determined not to do anything to hasten his death."[42] Such translations enable these contemporary scholars the possibility to entertain a broader range of interpretations of the story than other translations might allow. Indeed, these scholars generally do not align their arguments precisely by this *mutav* statement alone, but in coordination with other elements of the story.

Just before we move on to those other narrative details, note that all of the above bioethical renderings of Chananya's *mutav* statement suppress a critical fact. Chananya was put on the fire for a particular reason: he had committed a crime for which the government decided death was the appropriate punishment. Ascertaining why modern bioethicists do not acknowledge but actively disregard this larger political context may be an interesting question, but it would take us too far afield here. To be sure, given this larger political context, it would be easier to understand Chananya's *mutav* as an expression of acceptance of the authority's judgment against him, an acceptance he earlier and elsewhere articulated with biblical scripture—than as some bioethical principle regarding intervening in someone else's dying processes. "It is better that the one who takes it is the one who gave it. One should not injure oneself"—perhaps functions as a theo-political principle: insofar as the body in its totality belongs to God, one should resist being complicit with a human government's morbid or mortal injury of God's property. When he utters this phrase, Chananya is not suffering the end stages of a particular disease; he is enduring capital punishment. It therefore remains an open question why bioethicists fixate on—and fix—this apparently *political* statement to buttress their *bioethical* arguments.

Executioner

Though some bioethicists stop their reading of the story here at Chananya's *mutav* statement and hinge their positions vis-à-vis euthanasia based on their truncated rehearsal of the *sugya*, many more continue the narrative to the next encounter.[43] Here (paragraph 6 above) the executioner speaks with Chananya as he burns. He offers Chananya a deal:

> "Teacher, if I increase the flames and remove the tufts of wool from your heart, will you bring me to the World to Come?" He said to him, "Yes." "Swear to me." He swore to him. Thereupon he immediately increased the flames and removed the tufts of wool from his heart. His soul quickly departed. He even jumped and fell into the fire.

We will return to Chananya's response to the executioner's plan in a moment. For now, we examine what the executioner offers to, and does for, Chananya and why—and how modern bioethicists interpret these details.

Some bioethicists misconstrue the executioner's question. While it is true that the executioner asks an inverted direct binary question—will Chananya bring him into the World to Come—the answer to which is either yes or no, Rosner reframes the question to be even more self-serving than what is recorded in the Talmud: "Rabbi, if I raise the flame and remove the tufts of wet wool from your heart, will I enter the world-to-come?"[44] This version of the question both erases and precludes Chananya's complicity to the executioner's plan. This is a curious move for Rosner, however. Just two paragraphs later in his essay Rosner recites a different story of a Roman executioner asking R. Gamliel, "If I save you, will you bring me into the world-to-come"—a translation that both keeps true to the Talmudic text and retains Gamliel as an active participant in what happens to him.[45] So why does Rosner remove Chananya's agency here? A partial answer can be found in the Talmudic dictum Rosner invokes further on, that "no law (*halakhah*) may be quoted in the name of one who surrenders himself to meet death for the words of the Torah."[46] Perhaps Rosner alters the executioner's question so as to protect Chananya from being associated with any connotation of "surrendering himself" to his own death. In this way the story in general and Chananya's position therein can be invoked as an internally consistent support for Rosner's adamant argument that "suicide is a criminal act and strictly forbidden by Jewish law."[47] Had he quoted the *Avodah Zarah* text as it actually is, Rosner would not be able to point to it as foundational support for his desired conclusion.

Another common reading strategy regarding the executioner suppresses his motivation. Though it is clear in the text he desires access to the World to Come, some bioethicists overlook this detail altogether so as to highlight other dimensions of this exchange.[48] For example, according to Sherwin, "The Roman executioner asks the rabbi whether he may remove the tufts of wool from over his heart that artificially prolong his life."[49] This is inaccurate because it is incomplete. The executioner asks Chananya what Chananya would or could do were he to intervene in his flaming death, which is precisely this: bring him, the executioner, into the World to Come. His was not a question intended for Chananya's benefit; it was self-serving. Green countermands the text by suppressing the executioner's motivation and portraying him as a compassionate individual. "At this point, we read that one of the Roman executioners, taking pity on the rabbi, 'raised the flame and removed the tufts of wool from over his heart', permitting the rabbi's soul to depart. The narrative goes on to tell us that because of this act of mercy, the executioner 'went straight to heaven when he died.' "[50] While it would be nice to imagine the executioner interacting with Chananya with some semblance of pity, mercy, even compassion—this is not what *Avodah Zarah* conveys. As we noted in Chapter 3, such a kind executioner does exist in the textual tradition, but only in the narrative in BT *Kallah* 51b. Rather, our text is explicit that the executioner here is primarily if not exclusively self-interested. He inquires whether he can get what he wants from Chananya; he does not investigate how he can comfort the burning sage or mitigate his undeniable suffering for Chananya's sake. No doubt suppressing—erasing—the executioner's self-serving motivations strengthens Sherwin's and Green's argument that since heaven bestowed approval upon this supposedly benign executioner, the text as a whole endorses, "as morally praiseworthy at least some lethal interventions aimed at putting an end to a dying person's suffering."[51]

Would it have only complicated things or undermined their fairly permissive positions were they to acknowledge the executioner's actual motivation? Perhaps not. According to Green's and Sherwin's rosy lenses, the executioner asks and acts for the dying's benefit—for which he gets a reward; hence, it is reasonable to say that the story generally supports the idea and practice of intervening in a dying person's demise. Yet consider the alternative, more complete and accurate, reading: acknowledging the executioner's narcissistic motivations and the fact that he received exactly the same reward as the hypothetical magnanimous executioner—easily supports the notion that *whatever* one's motivation for intervening in a dying person's demise, divine approbation may be possible. For surely were one to be magnanimous as these bioethicists imagine, heavenly reward would await.

Yet the text does not say this; it says reward is possible even for one whose motivations are self-serving. One need not be a saint to remove impediments to or hasten a death.

Another hermeneutical strategy brought to bear on this exchange involves denial and deference. Bleich acknowledges that it is plausible to read the executioner's offer to remove the tufts of wool as permission to withdraw impediments to death. Yet, this argument, Bleich asserts, "is based upon a misreading of the text. As recorded in the Gemara, the words of the executioner were: 'Rabbi, if I *increase the flame* (emphasis added) and take away the tufts of wool from over your heart, will you bring me to life in the world-to-come?' "[52] Bleich's careful attention to detail of this particular exchange enables him to describe that the executioner "engaged in active euthanasia," yet he immediately declares this "an act that is categorically prohibited."[53] It appears that Bleich does not forge ahead in the narrative to the *sugya*'s conclusion where the executioner receives divine imprimatur, which, for me at least, precludes claiming that actively hastening an individuals' death is "categorically prohibited." Still, Bleich demonstrates how to read carefully, albeit selectively. By cutting the story short he denies the text its own possible normative conclusion. This detailed yet truncated reading of the story apparently gives him sufficient legal room to declare by fiat that active euthanasia is prohibited. Yet, insofar as his is an abbreviated consideration of this classic source, how much deference to his own declaration is reasonable? May we readers of his work stop short of its end?

On this point, Steinberg also reads this exchange between the executioner and Chananya in a wholesome way, yet he abandons it in favor of deferring to another authority. He describes the executioner's action as "active mercy killing" because "he actively hastened the death" of Chananya "because he was a 'terminally ill patient' who was suffering terrible agony," and for this act "the Roman earned himself a place in the World to Come."[54] We will investigate below how and why it might be reasonable to consider Chananya a patient. Yet Steinberg does not conclude his argument here. He immediately invokes Moshe Feinstein's position that "this was an 'emergency' decision (*hora'at sha'ah*), and does not represent a general principle which we may apply.... The halakhic authorities conclude—unanimously, I believe—that there is no circumstance in which it is halakhically permissible to hasten anyone's death."[55] Both legally and logically this argument of deferral is ultimately unreasonable. On the one hand, it denies the very nature of emergency declarations. These are temporary, ad hoc rulings peculiar to specific geo-temporal circumstances, and they cannot by definition become established as enduring norms for all times and places.[56] Were they to become standard rules, they would no longer be considered exceptional, momentary diversions from the typical.[57] Feinstein asserts that this story should be considered exceptional and not normative. On the other hand the story as it is in *Avodah Zarah* contains several norm-generating elements, including Chananya's generalized principle taught to his students and the *Bat Kol*'s articulation of divine reward to the executioner for his part in bringing about Chananya's death. These suggest that the text deserves to be read as normative, and that it does not describe the exceptional even though it is about an idiosyncratic circumstance. It is therefore unclear how and why Feinstein declares by fiat that this story must never be understood as having enduring normative status.

And there is the logic problem. Steinberg claims that there is unanimity among authorities that hastening another's death is impermissible. If this were the case, then the story—which he shows he has read to its conclusion—would not have the *Bat Kol* rewarding the executioner. Indeed, no version of the story in which the executioner receives this end would have been recorded had this unanimity truly existed. But since the *Bat Kol* in both this story and those other stories does exist and these stories have been variously invoked by rabbis through the centuries, it is illogical to claim that no authority—heavenly or humanly—supports the notion or practice of hastening another's death.

Some bioethicists abstract from this exchange altogether. In lieu of reciting this exchange as it is, Novak silences both the executioner and Chananya: "The Roman official charged with killing Rabbi Hanina does shorten his death-agony, which seems to have been his right."[58] But this contradicts his position we noted earlier, that "someone who has political power or authority over some other human or humans has no right to destroy or command others to destroy that human life, even if that 'other' is the victim himself or herself."[59] It seems Novak does not take into consideration the *Bat Kol*'s imprimatur of the executioner's action; apparently heaven indeed endorses the idea and practice of a human authority taking another human's life. To be sure, Chananya does not give his assent to his own demise per se; that was already a foregone conclusion when his punishment was announced, though at that moment he praised the justness of the judgment. Rather, what Chananya agrees to in this conversation with the executioner is *how* his death should come about, not *whether* it should. The *Bat Kol*'s commendation of Chananya thus apparently supports the notion that it is permissible—or at least commendable post facto—to grant permission to an authority to hasten one's own death, especially if one's death is already inevitable *and* imminent.

Each of these readings of the exchange between the executioner and Chananya has its strengths insofar as each enables a bioethicist to say the text supports certain claims and conclusions. Yet every reading strategy reveals critical weaknesses. Whether it corrupts the story by emending or eliding a character's agency or motivation, or it denies the text its normative elements and status, or it glosses the text in such a way that it appears to speak about something it does not—such reading strategies undermine the integrity of the text as well as the contemporary bioethical arguments in which they are found.

Chananya's response

Despite and perhaps because of these difficulties of looking to the executioner for normative guidance regarding euthanasia, many bioethicists focus more on the dying individual himself. For them, Chananya is the more important character and he should be the source for any subsequent norm regarding intervening in someone's dying processes. For purposes of clarification we should recall what it was that Chananya said and did vis-à-vis the executioner (this is again paragraph 6 above):

> The executioner said to him, "Teacher, if I increase the flames and remove the tufts of wool from your heart, will you bring me to the World to Come?" He said to him, "Yes." "Swear to me." He swore to him. Thereupon he immediately increased the flames and removed the tufts of wool from his heart. His soul quickly departed. He even jumped and fell into the fire.

Contemporary bioethicists read Chananya's response to the executioner in two distinct ways. The first method alters what it is that Chananya agrees to in the first place. The second method takes into consideration the larger context in which this particular exchange occurs, and in so doing, finds complications.

Those employing the first method invariably suppress the fact that the executioner increases the flames surrounding Chananya. Rather what matters for these readers is that Chananya agrees to the removal of the wet tufts of wool. Their removal constitutes a textual precedent for passive euthanasia insofar as these tufts are "artificial" and thus akin to technologies that prolong life.[60] Whereas most scholars maintain that Chananya agreed verbally to this proposed set of actions, some do not. Dorff, for example, contends that he "allowed his students to bribe the executioner to detach them," and Michael Barilan suggests that Chananya paid the executioner to remove the tufts.[61] Though such readings lack textual support, all these scholars agree that Chananya's assent to the executioner's plan to remove the tufts of wool serves as a form of passive euthanasia.

Not all bioethicists are content with this claim, however. Shulman, for one, offers a convoluted reading that focuses on the removal of the tufts of wool. But he does so by conflating Chananya's response to the executioner with Chananya's response to his students. According to Shulman, Chananya did "allow the executioner to remove the wet sponges; the fire could then consume at its natural unimpeded pace. This act by the executioner of removing hindrances to natural death was deemed meritorious. The conclusion is that a terminally ill patient may refuse certain treatments in order that his agony not be prolonged. Rabbi Hananya refused to open his mouth, because suicide and euthanasia are forbidden."[62] Chananya does not refuse the "treatments" the executioner proposes to remove, nor could he refuse them now: they were *already* attached to him much earlier in the narrative. Rather, the most that Chananya can do—and that he does do—is grant permission for their removal now. *Allowing* intervention is rhetorically and authoritatively different from *refusing* treatment. And, to be clear, Chananya's only instance of refusal is toward his students' proposal. That refusal prolongs his agony—precisely contradicting Shulman's conclusion. In this way and perhaps unintentionally, Shulman's argument demonstrates how narratives offer both foundations for norms and subversions of those norms, just as Adler and Cover taught (see Chapter 2).

Regarding Chananya's acceptance of the executioner's proposal, Shulman states, "many interpret this source to teach that just as the sage allowed the wool to be removed, so is it permissible for a man whose life is of unacceptable quality to him to request that treatment be stopped. Withdrawing all therapy would be a proper treatment for an individual who wants to die, when he is suffering intractable pain. Consequently, a terminally ill patient may request withholding of therapy in order that his agony not be prolonged."[63] Where did Chananya *request* anything? According to the text Shulman consults (*Avodah Zarah*), Chananya is a passive recipient of others' ideas: his students suggest self-inflicted asphyxiation, his executioner proposes a two-fold intervention. The former Chananya rejects, the latter he accepts. Yet in neither instance does he creatively propose what should, or request what could, happen to him while he dies. Shulman reads into the story what he wants to see therein.

Rosner also uses exactly the same translation of this section as Shulman that highlights Chananya's permission to remove the tufts of wool.[64] This leads Rosner, like many other scholars here, to permit withdrawing treatments from dying patients in certain circumstances. Yet Rosner, as do all these other scholars, suppresses the fact that Chananya assents to the totality of the executioner's proposal, not just a portion of it. He agrees to the plan that incorporates *both* the removal of the wet tufts of wool *and* the intensification of the flames. The text, including all the manuscript variants, insists that Chananya agrees to the whole deal. He does not say yes to one line of proposed action and no to the other. Why is it that these bioethicists are silent about this fact? Why do they contort this exchange? And for those few who do acknowledge the proposal to increase the flames, why do they interpret this to mean that the fire would consume Chananya at its "natural, unimpeded pace" and not at a more rapid rate that intensified flames surely would produce? And this is where we encounter the second method of reading Chananya's reaction to the executioner's proposal.

One possible answer is that it might lend support for arguments promoting active euthanasia. Weiner, for example, cites both facets of the executioner's proposal. Weiner explicitly states that the removal of the tufts of wool could support the notion of permitting passive euthanasia, yet immediately opines that this "analysis constitutes a misinterpretation of the events. The Rambam states that not only is a direct act of killing defined as murder, but also [so is a] death caused passively (e.g., ... the Roman guard not only removed the sponges but also made the fire greater [an act that directly hastened the rabbi's death])."[65] For Weiner, increasing the flames is an act of active euthanasia and is as prohibited as is passive euthanasia, a curious claim that counters what most other bioethicists see in this story.

Weiner complicates things further when later on he brings in the *Kallah* version to augment his position. "It was this Roman guard who originally lit the fire. His being the murderer of R. Chanina enabled R. Chanina to agree that he should at least do it more 'humanely'. (Hastening death is clearly not considered a new act of murder.)"[66] As shown in Chapter 3, the *Kallah* version of the story does not support Weiner's claim that the executioner actually lit the fire underneath Chananya; rather, the executioner witnesses the fire retreat from Chananya, exclaims wonder about this unnatural occurrence, and offers to release Chananya altogether from the flames. Undeterred by these facts of the texts, Weiner conflates the *Kallah* version with the *Avodah Zarah* one, using the one to justify Chananya's actions in the other. But if both passive and active euthanasia are *ab initio* prohibited, then why bother seeking another text that apparently justifies Chananya's consent to them both? And why grant the *Kallah* version such normative authority when the *Avodah Zarah* probably suffices for his purposes?

Nevertheless, Weiner plunges on toward another line of argumentation. Why, he wonders, did Chananya agree to hasten his death? Instead, "he should have preferred suffering as long as there was any possibility of *chayey sha'ah*."[67] *Chayey sha'ah,* literally "a living hour," refers to someone living whose life is mortally imperiled; typical examples are those who have been crushed in fallen buildings, and for them others are to desecrate Shabbat by digging through rubble in order to locate and extract them.[68] As long as there was some possibility of others intervening to save him, Chananya should have done everything in his power to stay alive. But this is not how Weiner understands the situation. "In R. Chanina's case, his obligation was to give up his life; sacrificing his *chayay sha'ah* was also a fulfillment of '*with your entire soul*' [Deuteronomy 6:5]. Therefore, he was able to agree to hasten his own death through the Roman guard. (But for himself to hasten his own death, even in a case of *Kiddush Hashem* [martyrdom], would have been prohibited.)"[69] This is confusing, to say the least. At one level Chananya should have rejected the executioner's proposal so as to stay alive even for a few more blistering moments on the tiny chance that others might save him; at another level Chananya agrees to hasten his own death with the executioner functioning as his deputy; and at a third level Chananya fulfills the three-fold biblical commandment to love God with all one's heart (*lev*), soul (*nefesh*), and might (*me'od*) (see Deutereonomy 6:5) and in the process forfeits the potential salvation intimated in the *chayay sha'ah.* If it were truly the case that Chananya was "obliged" to give up his life, saying that he "was able to agree" to hasten his death, is vacuous: he could not have done otherwise. And saying that he should have remained alive and not agree to the executioner's plans is to suggest that he not fulfill the biblical command to die "with his entire soul" (if that is applicable in the first place). And Weiner's final parenthetical comment assumes that there is a substantive, qualitative difference between verbally agreeing for another to hasten one's death and personally hastening one's death. If hastening another's death—actively or passively—is as wrong as Weiner wants us to think it is, then his assumption is thwarted by the well-known Talmudic dictum that one may not deputize another to do a transgression.[70] At one and the same time, Weiner justifies and undermines Chananya's assent to the executioner's plan, and he condemns and condones hastening death (whether one's own or another, it is unclear). Despite and because of these frustrating problems in Weiner's reading of the story, Weiner's argument demonstrates why this *sugya* is so fascinating, as it can be read in so many ways simultaneously.

Other bioethicists labor alongside Weiner in trying to understand Chananya's response to the executioner's proposal. Telushkin, for example, states simply, "But, as his situation became more and more acute, with death inevitable, he finally concluded that it would be best to die as quickly as possible."[71] Chananya accedes to the executioner's plans because he sees no alternative, no salvation, no mitigation. With no way out, Chananya did what was best in his situation: agree to the executioner's two-fold plan. Telushkin then offers a visceral contemporary version of this story: "An analogy can be drawn with those individuals who jumped out of the high floors in the Twin Towers on 9/11. At first, when the people inside the towers started to feel the fire's heat, they tried to escape or to find some way to hang on for as long as possible. But when the fire started to create unbearable heat, and both escape and rescue seemed impossible, some people chose to leap to certain death if only to put an end to their suffering."[72] Though vivid, this analogy suffers a critical fact: there was no executioner offering these victims a deal. Whereas Chananya was given a choice by the executioner—to die longer or quicker, these victims were not given such a gift; they created their own way out—literally and figuratively.

Telushkin nevertheless points to a critical element of the story that the vast majority of bioethicists ignore completely. Time passes. Characters change with the passage of time. And Chananya is no exception. Kravitz, too, latches on to this temporal dimension—about which I will say more below.

For now, Kravitz says this about Chananya's reaction to the executioner. "As given in the story, the kindly executioner's question was more than a request for information; it was a proposal of a contract. That the executioner asked the rabbi to swear to his answer indicates that both executioner and rabbi knew what was to be the outcome and what was to be the consideration, the quid pro quo."[73] Why Kravitz portrays the executioner as "kindly" is unclear. But he is clear that what the executioner develops is a business plan, a contract, between him and Chananya. He wanted something (access to the World to Come) and he surmised that Chananya wanted something (access to relief from his agony). He could supply the latter if Chananya could promise the former. Quid pro quo. Kravitz is therefore correct in saying that Chananya agreed to a contract. A contract entails at least these elements: clear temporal boundaries, negotiated conditions, termination clauses, and the parties thereto are relatively equal, at least in terms of the contract.[74] Certainly the exchange between the executioner and Chananya meets these criteria: their contract will begin as soon as both assent to it and it will end upon the executioner accessing the World to Come; they both agree to the content of this exchange; it will be void if either fails—willfully or accidentally; and both enter into the contract of their own accord, that is, as relative equals. Perhaps enabling this relative equality to come about is what prompts Kravitz to see the executioner in a positive light.

To the End

There can be no doubt that in Jewish bioethical literature the three most referenced elements of this narrative are Chananya's response to his students (the *mutav* statement), the executioner's proposal, and Chananya's acceptance of that proposal. Just a few bioethicists carry on, however, and read to the *sugya*'s conclusion. There they encounter the *Bat Kol,* the heavenly voice, which declares that both the executioner and Chananya have been assigned to the World to Come. The fact that the executioner achieves exactly the same reward as Chananya suggests that at least some level of heavenly approval for the executioner exists. Now it could be that he receives this eternal reward as just desserts for the totality of his life; that is, he being the executioner in this particular instance might have been discounted. Or it could be that his actions with this particular victim of capital punishment were sufficient cause to justify the bounteous reward. Though the *Bat Kol* is vague on this point, Rabbi Judah HaNasi is not. For him, the executioner was otherwise an unworthy gentleman; only how he treated Chananya in this particular circumstance earned him access to the World to Come. His recent actions—not his life in general—were meritorious.

The *sugya* could have stopped with the *Bat Kol,* leaving the contemporary reader to discern why the executioner warranted heaven's praise. But with Rabbi Judah HaNasi's comment included and reinscribed in every manuscript of *Avodah Zarah,* it seems that the shapers of the rabbinic tradition want readers of this narrative to see the executioner's actions in and of themselves as meritorious, not just magnanimous. Perhaps it is recorded in this manner to echo the principle taught elsewhere by R. Joshua that the righteous of the nations of the world will have a share in the World to Come.[75] Put in contemporary terms, the Talmud primes the modern reader to see the executioner's actions positively.

It should therefore not be surprising when some bioethicists praise the executioner as well. Peter Knobel, for one, interprets the granting of eternal life to the executioner this way: "In fact one can read this passage to suggest that relief of suffering which hastens death is not only permitted but meritorious, so meritorious that the executioner is immediately ushered into eternal life."[76] This means that for Knobel assisted suicide is permissible, if assisted suicide means, "the person is unable to act for certain emotional or moral reasons but is able to permit another to help him." In this way he distinguishes verbal from physical action, whereas Weiner does not. Green also takes the narrative's conclusion seriously when he says the story "boldly upholds efforts to hasten death in at least two different ways: first, by removing impediments that slow a painful and certain dying process (the damp tufts of wool), and, second, by actively taking measures to accelerate dying (raising the flames)."[77] The inclusion of the *Bat* Kol in the story serves as undeniable proof to Green and others that "the classical sages accepted as morally praiseworthy at least some lethal interventions aimed at putting an end to a dying person's suffering."[78] Given such classic support for intervention, Green holds that "the relief of suffering can sometimes take priority over the protection or continuance of biological human existence."[79] Kravitz pushes the envelope further. For him the story's conclusion is not merely permissive of intervening in someone's intractable suffering, but it also points to the necessity for such intervention. "Where pain trumps life, where suffering cannot be controlled and recovery cannot be achieved, then if the patient feels that life is no longer worth living, and 'the game not worth the candle', there is no need to extend life, and indeed, there may be a need to shorten it."[80]

This is a controversial claim, certainly. Excruciating pain may justify intervention for some bioethicists, but others insist that even then no one may interfere. Prouser says Chananya "taught us through both deed and dictum [that suicide is forbidden] even for one suffering pain and illness."[81] For Herring the story "provides an authoritative precedent to deny permission to hasten one's death by an overt act (such as opening one's mouth to swallow the flames surrounding one), even when confronted with the prospect of great pain preceding death."[82] A painful life, however meager it may be, however short its prognosis, is better than death.[83] Indeed, suffering itself may be therapeutic for the individual insofar as it allows time and motive for the individual to reevaluate his or her deeds and return to God.[84] Though such interpretations may lead to lengthening a particular person's (painful) life, they nevertheless cut short the story itself because they cut out—eviscerate, if you will—details critical to the narrative's richness, to its very vitality.

Killing a Dying Story

All these reading strategies pervade contemporary Jewish bioethical conversations on euthanasia. No particular school of thought or single stream of modern Jewry holds the monopoly on them. That said, as demonstrated above, some bioethicists are quite attentive to what the text actually says, conforming their interpretations and rulings to what can reasonably be grounded in the version of the story they cite. Many more, however, are not so careful. Most bioethicists run roughshod over this text. They cut, contort, silence, gloss, and sew it together so that it "says" what they want it to say. In so doing they kill the story, to be sure, but perhaps they do more damage than just that.

One feature of the story is especially vulnerable. It is the dimension of time. As noted at this chapter's beginning, every narrative necessarily involves characters in a particular context, as well as time and change. Every version of Chananya's fiery death entails these elements and the *Avodah Zarah* tale exemplifies them most clearly. This fact is lost on most bioethicists, however. Temporally speaking, the text is flat or thin for them. What they seek in the story is its conclusion, not its development. They plumb the story for its normative position on the topic of their concern and they discard the rest. They are not interested in the characters embedded in the narrative as personages in and of themselves. They do not seem to care about the larger context in which these characters exist. Suppressing all these features in favor of honing in on what they want to see denies the characters their very nature as human beings who exist in and through time and also who change based on their experiences in and through time.

In their eagerness to see in the story a singular and consistent position vis-à-vis end-of-life issues such as suicide and euthanasia, bioethicists impose upon the text what it cannot support. Kravtiz, perhaps the most outspoken scholar on this point, has this to say: "The story does not present a consistent position. Since it purports to describe events occurring over time, the story suggests that Rabbi Hananiah's immense suffering led him to change his mind and facilitate his own death!"[85] Kravitz understandably focuses on Chananya since he is the story's protagonist. He sees in Chananya a change of mind sparked, as it were, by his experience of enduring the flames. Earlier on in his experience, Chananya expresses to his students his reticence to injure himself. But the story—and the flames—do not end with that principled instruction. It is after he finishes teaching his students one last time that the executioner approaches him with his proposal. How much time passes between these two exchanges? A moment? A minute? An hour? An afternoon? The text does not say. Though the quantity may not matter, we can be certain that *some* time elapses between these two moments of the story. It is during and through that time that Chananya reassess his situation. His external circumstances have not changed: the fire still rages with no end in sight; his body still fuels and feels the fire; his chances of leaving the flames alive remain nil.

If the outside world has not changed, what has? What evidence is there that anything has changed at all? About this the text is abundantly clear: Chananya assents to the executioner's plan. Whereas earlier he refused to harm himself further than he already had been, now he agrees to an action that would cause him terminal harm. Something changes within Chananya. He changes his mind. His experience and his assessment of it lead him to a position different than the one he held earlier. Sentient and cogitating creatures do precisely this in their pursuit to exist and thrive. They learn from their experiences so as to continue living. Living requires changing. In his very dying, Chananya evidences what it means to live. Paradoxically, he is only now dying to die.

This shift is profound, yet modern bioethicists apparently balk at this fact, as proven by their general disregard for Chananya's inconsistency. Were Chananya consistent as bioethicists want him to be, he would have responded to the executioner either with silence or with this: "If you are asking for information, that is one thing, but if you asking for my concurrence, that is another, for I will remain constant in position and consistent with my earlier statement, and therefore will do nothing which might in any way speed my death."[86] Either response—total silence or articulated silence—would justify seeing in Chananya a consistently held conviction that agreeing to harm oneself or have harm done to oneself is improper. Yet such silence and consistency are not what the *sugya* offers.

Expecting Chananya to be consistent despite his experience of excruciating suffering in and through time would be unreasonable. And forcing the text to be read that way would be equally problematic if not cruel. Again, Kravitz: "It would have been inhuman to expect him to follow either alternative. It is human, understandable, and, indeed, to be expected that Hananiah acted as he did; faced by certain death and experiencing terrible pain, he sought to avoid the latter by accelerating the former."[87] Kravitz's counterfactual thought experiment highlights the problem bioethicists face when seeking consistency in Chananya.

What Kravitz fails to do, however, is note that other characters are involved as well. His daughter is a ready example of another character that undergoes change through time. Recall that before Chananya is wrapped in the Torah scroll and strapped to the pyre, he and his wife and his daughter emerge from the tribunal and exclaim the justness of the verdicts rendered upon them. Now as she sees her father aflame, she is understandably distressed. She may not explicitly question the justness of the verdict against her father. But she certainly expresses the anxiety of seeing her beloved in such agony. Even if one were to read this *sugya* without knowing the daughter's pious response to the punishment, her reaction here to her father's plight would no doubt be understandable. Many a daughter would be distraught at seeing her father in such pain. But knowing her as a character with a history prior to this particular moment enriches our appreciation of her not just as a(ny) daughter but as Chananya's religiously inspired daughter. For knowing her earlier piety and solidarity enables us to see in her conversation with her father the profound depths of her theological anguish. That background renders all the more sensible Chananya's theological response to her; without knowing this historical context, without taking it into consideration, his words to her seem extraordinarily pious and therefore, in the eyes of most bioethicists, liable to be dismissed. And yet, now that we recall her in her totality, his last words to her reflect his intimate knowledge of his daughter: he sees in her a similar theological bent as his own. Consoling her just now will best be achieved through theological rhetoric—something only a father would know because he has known her for a long time. Just as Chananya sees change in his daughter through time that is sparked by his *crematio*, should we not also see change in him?

Taking Kravitz's point even further, consistency is hard to find were we to take the narrative *as a whole*. To his daughter Chananya expresses theological comfort, that he accepts his circumstance and there is no need to intervene one way or another in his dying moments. If to her he says intervention is not necessary, to his students he says something entirely different: intervention, be it by himself or another, is impermissible. Toward his students he articulates a clear and strongly held position. Yet, in his assent to the executioner's proposal, Chananya demonstrates a third position, one that is strongly in favor of intervention.

Three distinct positions about intervening in his dying are readily evident in the *sugya*: relative indifference, negative, and positive. And these positions occur precisely when they do as the story unfolds and as the characters change. Had the story incorporated only one of those moments, it perhaps would not have fascinated subsequent readers. It holds our attention for the very fact that it traces characters who change in and through their experiences. Replete with their own histories, emotions, and motivations, these characters are realistic. They live—and change—in and through their experiences. Saying the *sugya* holds a consistent position, as do so many bioethicists, betrays the text and ignores the very humanity of the characters invoked therein.

Flattening the temporal dimension of the story is but one way bioethicists impose upon the text. Another way is to assume its relevance in the first place. The vast majority of bioethicists assume that this particular story both speaks about euthanasia generally and has relevance to contemporary instantiations of euthanasia in particular. This assumption may be misguided, however. As Kravitz notes, modern treatments of euthanasia among Jewish bioethicists lean heavily on just a few classical texts. And these texts, like the one examined here, lend themselves to multiple interpretations. Moreover, "It is assumed that the texts describe real events and that the conclusions derived from them are germane to our present situation. They may not be."[88] That is, these texts may not speak about real events *and* they may not be relevant to contemporary euthanasia.

The plethora of stories regarding Chananya's demise supports the first suspicion. Insofar as all these texts exist and offer vivid disagreements about what happened to Chananya, who was around, what he said, and what ultimately happened—questions arise as to which version is accurate. The best we can do is surmise, or appoint, one as the most accurate. Obviously, this cannot provide us with veracity. The historicity of these texts remains obscure, and to claim otherwise is intellectually questionable.

As for relevance, this question is raised from both within and outside the text, from content and context. In regard to the former, Dorff acknowledges that though the story suggests the distinction between sustaining a person's life and prolonging the process of dying, its relevance to contemporary euthanasia deliberations is undermined by the fact that it "is not in a medical context."[89] Newman concurs, to a point. After reciting the story up to the moment of Chananya dying at the hands of the executioner, he says, "The conclusion of the story, while curious in a number of respects, appears to add nothing to the discussion of euthanasia and so is generally not cited in contemporary treatments of the issue."[90] Such reasoning would mean that the textual tradition's extension of heavenly welcome to the executioner—as well as to Chananya himself—should have no sway for contemporary bioethical thinkers. And it would silence Rabbi Judah HaNasi's lament that, in its own way, articulates ongoing rabbinic angst about interfering in end-of-life care. Squelching the concluding bits of the story snuffs out theological and social dimensions that are nonetheless intertwined with the care of a dying individual. For these reasons alone, the conclusion of the story—even if it does not conclude the way one would want it to—is relevant to bioethics. Nevertheless, this line of reasoning would say that since other classic sources have more relevance because they speak explicitly about medical contexts, they, and not this story—in whole or in part—should carry more weight in modern bioethical discourse. This story's bioethical normative relevance is compromised from within.[91]

And then there is the argument from the outside, from context, that undermines the relevance of this story for contemporary bioethical concerns. This is well illustrated in one of the earliest considerations of this topic by American Jews. At the 1950 gathering of the Central Conference of American Rabbis, R. Israel Bettan offered an initial foray into euthanasia in which the story of Chananya appears as the first rabbinic text cited, though he references the story only as far as the *mutav* statement. R. Jonah B. Wise rose to reply to the proposal, saying, "The question of euthanasia today is not one that can be discussed on the basis of the opinion of one who lived in Smyrna in the seventeenth century [R. Chayim Pallagi, referenced earlier by R. Solomon Freehof] or of our distinguished Rabbinical predecessors in Talmudic times. The moral question involved has, of course, been discussed by Dr. Bettan, but the world has progressed since that time; conditions have changed."[92] On the one hand, Wise assumes that the stories and texts emerging from ancient, medieval, premodern, and early modern periods were meant for the audiences populating those eras—and only those populations that shared those geopolitical and technological contexts.

On the other hand, this very assumption suggests a profound difference if not rupture between those periods and contemporary times. The modern context is so radically different from earlier ones, Wise seems to say, that it is unreasonable to invoke prior texts to shape modern norms because the authors of these texts themselves would have ruled differently were they living today.

Indeed, "had these Rabbis been aware of the circumstances that confront us, they would have changed their attitude." Wise's approach, however, faces a crucial challenge: if modern Jews are not to turn to earlier texts for normative guidance because the contextual differences are so dramatic, calling on those texts would be all but *verboten*. Since modern Jewish scholars would have few if any textual foundations to draw from, in which ways and to what degree could their ethical arguments and normative conclusions be considered Jewish? Wise responds to this challenge by encouraging his colleagues to consider those texts and contexts that speak of euthanasia explicitly, not merely implicitly.

This hesitance to turn to classic sources continues to exist, even half a century later. In his 2006 essay on the topic, Kravitz laments the paucity of texts speaking explicitly about euthanasia, and the few that are invoked are so vague that they can endorse multiple interpretations. Moreover, "It is assumed that the texts describe real events and that the conclusions derived from them are germane to our present situation. They may not be."[93] Even if one were to import this text into a bioethical milieu, difficulties remain, as Glick rightfully identifies: "It is not easy to translate any of these examples in a modern idiom. What are the analogies of the woodchopper, the salt, or the removal of the wet wool?"[94] For Glick and Kravitz as for Wise, relevance—both contextual and textual—matters more than textual authority.

The impulse to dismiss the story for internal and external reasons could be augmented by looking at it from another angle. As seen in Chapter 4, Chananya is perhaps most well known in the Jewish textual tradition as a martyr. He dies in part because of and for his beliefs. He engages in an illicit activity—teaching Torah publicly—and for this he receives capital punishment by the Roman authorities. Insisting his death is or should speak to contemporary bioethical deliberations on intervening in a patient's demise thus seems to be forced. It would impose an ancient political death story onto modern medical ones.

Even so, contemporary bioethicists should not so quickly throw the story aside. The Talmud itself links individuals slated for capital punishment and those in death throes (*goses*) because in both situations their cases are hopeless. Death is imminent for both, to be sure. Whereas it is permissible and condoned to mitigate the anxiety and suffering of the criminal condemned to death by ensuring he receives "a good death" (*mitah yafah*), the logic goes that so too a dying patient should receive similarly benign interventions that reduce suffering.[95] Pushing this analogy further, however, reveals some troubling inversions. The fire consuming Chananya from the outside—which is a penalty for his crime—is thus akin to a disease eating him from within. And the executioner, whose job description includes the singular task of killing captives, stands for physicians, whose tasks are to help patients secure relative levels of health, avoid pain, and, generally, to remain alive.

Despite these curiosities, such analogical arguments may salvage the story's relevance for bioethical treatments of euthanasia. Yet this very line of reasoning throws open a host of other issues. If stories of those condemned to death for criminal and political reasons are indeed comparable to imminently dying patients, then why should bioethicists limit themselves to this single classic martyrdom story? Why not also turn to other classic sources of martyrdom, some of which are treated in Chapter 4? And what about later cases of Jews being killed by other kinds of authorities for other reasons? But how reasonable is it to analogize a patient dying involuntarily from an illness to a criminal who willfully engaged in illicit activities that carry the death penalty as punishment? How would such patients and their families feel knowing that their care providers and Jewish bioethicists look upon them as no different than murderers? And, if they knew this particular story, how would they feel when Jewish bioethicists look more toward the Roman executioner for normative guidance than to a fellow Jew? What dangers lurk in leaning upon that ancient and self-centered *gentile* for establishing modern *Jewish* ethical norms?

One danger is the story's exceptionalism. As noted above, according to Steinberg, Moshe Feinstein rules that the executioner's "active mercy killing" "was an 'emergency' decision (*hora'at sha'ah*), and does not represent a general principle which we may apply."[96] Does this mean the centurion's decision to offer to intervene was an emergency decision, or that Chananya's agreement to the soldier's plan was? If the latter, it necessarily implies that Chananya took an active role hastening his own death, an action Steinberg claims is "patently forbidden under any circumstances according to halacha."[97] Were this claim taken seriously, it perforce means that the Roman centurion is the instigator and exemplar of this emergency decision. Again, this challenges the practice of hinging modern Jewish norms on the purported actions of a gentile whose sole responsibility was to kill Jews (among other people). But if we assume that it is Chananya's choices and actions that constitute the so-called emergency decision, if his demise is exceptional, how could it justify unexceptional norms for normal contemporary dying patients? The logic of exceptionalism undermines itself.

Another danger is the presumption that the story should be normative at all. Elsewhere the Talmud demonstrates that a *Bat Kol*, a divine voice, cannot and even should not override human interpretation of biblical texts and human norm creation.[98] In reference to Chananya's narrative, Novak therefore doubts "whether stories like this, especially involving heavenly voices, are actually prescriptive in a general sense."[99] This may be the underlying reason for the fact that so many bioethicists stop their readings of the story before they reach its end where the *Bat Kol* emerges. The very mention of the *Bat Kol*—regardless of what it actually says—would only compromise the normative value of the source text and thereby undermine the normative punch of their modern arguments. Cutting the narrative short solves this problem. Retaining the text's authoritative mien thus requires decapitating it—or amputating it, depending on how one views heavenly voices at the end of human stories.

Novak offers another reason why this story should perhaps not be viewed as normative: "Usually in the Talmud stories (*ma'aseh shehayah*) illustrate and specify some general prescription already stated."[100] There can be no doubt that this *sugya* is a story and does not render *p'sak halakhah,* a practical and unambiguous legal decision. Yet, it is far from clear what its general prescription is or what principle the story as a whole serves to specify. Both of these reasons urge a reevaluation of whether *any* story should be brought into modern bioethical discourse, and if so, how much normative valence should be ascribed to it. More will be said on these points in Chapter 6.

ENLIVENING A DEATH

For now the question is: might there be other ways to retain *this* story's relevance for contemporary bioethical deliberation on euthanasia without raising all these issues pertaining to its content, context, analogies, and voices?

No matter how it is done, making Chananya's *crematio* bioethically relevant will necessarily confront some of these challenges. Unfortunately, the ways in which most bioethicists invoke, quote, mistranslate, misconstrue, twist, cut, and contort the story—that is, the ways they eviscerate it—harm both this ancient story and perhaps modern patients, not to mention other stakeholders and bioethics itself. Though of course it should be assumed that these bioethicists do not mean to harm today's patients, it is difficult to hold this assumption in regard to how they treat the text itself. Time and again bioethicists configure the source text to say what they want it to say. If they want to promote the position that all forms of euthanasia are impermissible and bad, they usually read the story only so far (e.g., to the *mutav* statement). If they want to suggest that passive euthanasia is perhaps a plausible course of action, they might go on to read the executioner's plan to remove the wet tufts of wool, but would ignore and hence erase the plan to intensify the flames. If they want to put forward the idea that both passive and active euthanasia have warrant in the Judaic textual tradition, they might recite the executioner's plan in full. They read into the text what they want to see. However much this eisegesis may be convincing to certain audiences, it plunders the story and thereby does a disservice to the textual tradition in which it is found, and it helps put patients and families in situations they otherwise would not desire nor should choose.

Recall that the power asymmetry bioethicists enjoy vis-à-vis their audiences stems from their relatively greater familiarity with the textual tradition. It is rare that anyone would check the sources to ascertain the fidelity of their reportage when they say the textual tradition says something. Fellow bioethicists could serve as regulators and take each other to task for misusing classic sources. But this is more theoretical than real, because the vast majority of bioethicists also approach texts with their own agendas as the driving forces behind their hermeneutics. Drawing attention to such questionable strategies would probably bring undesired scrutiny upon one's own projects. So it goes on. With little to no oversight on their portrayals and interpretations of classic sources, bioethicists could get away with murder.

Though this may be alarming, it is not a call for bioethicists to turn in their pens and stop researching and writing on pressing moral biomedical issues. *Au contraire*. It is a call for bioethicists to be ever more vigilant about how they read classic sources generally and this ancient narrative in particular. It is a cry for care—for careful reading and cogent writing. It is a plea for taking the textual tradition seriously on its own terms, rather than imposing upon it one's own agenda. Instead of killing this story, enliven it by listening to it in all its ambiguity and ambivalence. Appreciate its narrative form and structure, the passage of time, and changes of mind. See its characters less as norm creators than as creatures anxious about the fragility of mortality.

It may be inevitable that incorporating this story for bioethical purposes may require selective reading. But selective reading can be done conscientiously. It can be tailored to meet the needs of the analogy. Instead of working backward from one's conclusion and reading the text in such a way that would support that position, this other approach considers the case at hand and looks at the source text for salient analogies. Here, the contemporary issue is euthanasia, the intervention in a dying patient's demise. So what and who in the story of Chananya's demise would be intelligible analogies to this case? Asking this type of question enables us to identify biomedically salient elements—and thus salvage the story's relevance writ large to the topic.

So if we are thinking about euthanasia, which details in this story would be bioethically salient and which peripheral? According to most other bioethicists, Chananya's last remarks to his students, the executioner's proposal and Chananya's response to it, are supposedly the most salient features. Yet each element has proven to be internally ambiguous. And the biomedical relevance of each is also questionable. Take, for example, the executioner. There is obviously some bioethical salience of the Roman centurion insofar as he is the one who actively intervenes in Chananya's dying process. Yet it behooves bioethicists to query how and why modern (Jewish) clinicians and care providers should shape their behavior like this (or any) executioner's actions. And if the executioner is so bioethically important, why do so many bioethicists deny the fact that the story includes heavenly approval of and human anxiety about his words and deeds? Selecting salient elements requires situating them in the fullness of the story, otherwise elements lose their details, nuance, and intrigue.

The bioethically salient details of this story would be those that make most sense when thinking about contemporary situations of possible euthanasia. One place to start is to consider who typically is involved in a dying patient's last moments. Who are the people typically making critical decisions regarding someone's final care? Generally speaking, family members are the most involved. Colleagues and students make rare appearances, and they are infrequently consulted about care decisions. Remember that Chananya's students show up only when he is already aflame for a while. Their tardiness reflects and reinforces the reality that students and colleagues may have ulterior motives guiding their interest in a particular person's demise. By contrast, his wife and daughter are present when he is sentenced, and his daughter is right there when he is first placed on the flames. She represents in so many ways the truth that adult children are typically the ones who care for and attend (to) the death of loved ones. Since this is the case, why not look at how Chananya's family interacts with him throughout the duration of his demise?

A second salient feature is the quality of the relationship between Chananya and the Roman centurion. Though theirs is a relationship between victim and executioner, it is not cut-throat, so to speak. The executioner takes Chananya seriously as a human being and not merely a corpse-in-waiting. He sees in his victim a person who deserves attention if not compassion, and he uses his role's authority to subvert existing power structures. And, most fascinatingly, he seeks a partner in this project. Whereas most bioethicists fixate on the portion of what the executioner proposes that corroborates their conclusion vis-à-vis euthanasia, I suggest that the very existence of his proposal and his repeated attempts to secure Chananya's assent to it are the more bioethically salient details.

Another significant feature to consider is the so-called patient himself as he endures his final moments. A closer and fuller look at him highlights changes in his attitude about death and intervention in his dying. His relationship with his own suffering also undergoes meaningful alterations. Such developments surely merit scrutiny in bioethical discourse because many if not most dying patients do not die instantaneously; rather, they perdure, adapt to their changing conditions, and often adjust their goals as they gain greater clarity of the severity of their circumstances.

A fourth salient feature is the character of the executioner himself. Perhaps bioethicists are right to look to his ideas and his actions—as he is the one who performs the deeds that intervene in Chananya's death. Yet, a closer inspection of his proposal and his motivations suggests that he is a complex character burdened with professional anxieties and theological goals. He thus appears more three-dimensional than most bioethicists construe him, and his very embodiment may provide insights about what it means to attend to the dying.

The salient features of this story to bioethical deliberations on euthanasia thus focus more on relationships than on norms per se. The specific relationships of concern here are family relations, authoritative relations, the patient's reflexive relations with his condition and prognosis, and the intervener's reflexive relations with his role and goals. The following few paragraphs put flesh on these features to demonstrate their pertinence.

Family Relations

Because family members and especially adult children typically make critical decisions for loved ones in their dying moments, it is only logical to study their roles in a dying story. Here we turn our attention to Chananya's wife and daughter. They stand in solidarity with Chananya when he receives his punishment from the otherwise absent Roman court. They emerge from that chamber simultaneously, and as one family they declare the justness of the verdict. Whatever reservations his wife and daughter may feel, their words betray none of it. They speak with the same verve and piety as he, as if to demonstrate to him and to the world that they willingly accompany him along this final journey. That they too are punished for his crimes reflects the justice system in ancient Roman-held jurisdictions. Could this not speak to and of the experience of many family members who feel that they too have been burdened, sentenced even, with a severe decree when they learn of a loved one's imminent demise?

Now recall the exchange between Chananya and his daughter as it occurs in our central story (paragraph 4 above):

> His daughter said to him, "Father, to see you thus!" He replied to her, "Were I to be burned alone it would be a difficult matter for me. Now as it is me who burns and the Torah scroll is with me, the one who will address the affront to the Torah scroll will also address the affront to me."

That Chananya's daughter appears at his side when he is set aflame seems to be purposive in the story. On the one side, she demonstrates the anguish of children as they endure their parents' demise, and on the other he offers her a dying gift. This very exchange suggests something important for caring for loved ones in their final moments.

Hers is a realistic cry about witnessing her father in such a state. (That his wife is absent at this point may be due to the fact that she has already been killed, or does not want to suffer witnessing her beloved husband's last and excruciating moments.) His response to her is that she should not worry so much about him, because, for better and for worse, he will die at some point soon. Nor should she concern herself so much with the burning Torah, the symbol of their cultural heritage, because culture is larger than this one scroll or this one human body. Their culture, he reassures her, perceives his early death as perverse and will respond to it in some unspecified way. I suggest that "the one who will address the affront to the Torah scroll" is culture and not God per se because Chananya does not mention God explicitly though he could have. Moreover, culture can transcend such earthly persecution; God is already impervious to such petty things. Chananya's and his daughter's culture, the very community in which they exist, takes umbrage at the insult imposed upon him and the Torah. Her comfort should be found there, in the fact that her community—and their God, to be sure—notices and cares about their family. Indeed, her identity intermingles with this larger community of people whose history she shares and with whom she will forge a common future. Trust this larger context, he urges.

His soothing words transmit to her a worldview replete with its values and people. He wants to guide her, even now in his dying and even tomorrow beyond his death. He communicates to her a kind of living will, a testament providing her a semblance of the moral guidance he has given her all his living days. By contextualizing his life and his dying, by framing it as he does, he empowers her to live on despite *and* because of his death.

She is able to receive these enlivening words only because she is physically present as he suffers his last. She has access to him, and he to her. In this final moment of earthly proximity, she offers him her daughterly affection and concern, and he in turn extends to her his paternal comfort and guidance. Could not similar opportunities be secured for families when someone is about to die? Providing family members unfettered access to dying patients may enable them to express to each other final sentiments that hitherto had not been communicated. Sidelining and silencing this interaction between Chananya and his daughter undermines the honor due to such relationships and curtails the importance of these moments for contemporary families who undoubtedly will be affected by the impending death.

Authoritative Relations

Another look at the executioner highlights different details that are, I argue, more bioethically salient than the two elements of his proposed actions that most bioethicists hone in on. For precision's sake, here is the exchange again (paragraph 6 above):

> The executioner said to him, "Teacher, if I increase the flames and remove the tufts of wool from your heart, will you bring me to the World to Come?" He said to him, "Yes." "Swear to me." He swore to him. Thereupon he immediately increased the flames and removed the tufts of wool from his heart. His soul quickly departed. He even jumped and fell into the fire.

First, that the executioner does anything out of the ordinary is astounding. He could have said nothing and let Chananya die according to the Roman regime's plans. Chananya could have been just another Jewish dissident executed for a crime and his death would not have made much of a splash in the Talmud, much less in contemporary bioethical discourse. But the executioner interrupts the flow of the flames, the hiss of justice. His very interjection thwarts not his victim but his employers; by speaking to this dying man he subverts Roman authority. By speaking at all he transforms himself from a thoughtless apparatchik to a thoughtful—albeit conflicted—person.

Now that we hear his voice crackling through the story, what does he say? Most bioethicists hear only his proposal to remove the wet tufts of wool. To their ears this means something akin to the withdrawal of life-sustaining treatments, technologies often associated with passive euthanasia strategies. They then wrestle with allocating the appropriate valence to this mention, whether it should be understood positively or negatively. As already seen above, some bioethicists read this mention in light of the *mutav* statement uttered to his students and therefore see the removal of wet tufts of wool negatively. Others take inspiration from the story's conclusion and impute to this strategy a level of positive endorsement. Yet the executioner speaks of more than wet tufts of wool.

He also suggests that he will increase the intensity of the flames now scalding Chananya. Some bioethicists read this as a proposal similar to the application of lethal amounts of sedatives or other drugs, techniques often associated with active euthanasia. Again bioethicists struggle to ascribe to this act a valence that comports with their ultimate conclusions, and it depends on how they read the *mutav* statement and whether they reach the story's end. (Some may wonder if the modern notion of *withholding* life-sustaining treatments can or even should have a textual foothold in this story. I suggest that it is unnecessary to rely upon this single story for each and every medical procedure employed in contemporary euthanasia cases. Furthermore, there are other classic texts that speak explicitly about the withholding of life-sustaining, or seemingly life-sustaining, treatments or features, so that an otherwise moribund individual can die.[101] Instead of relying on this particular story alone, we would do well to consult those other texts and meditate on them before ruling one way or another on withholding life-sustaining treatments.) And then the bioethicists move on, as if this is the totality of what the executioner said.

What the executioner actually says is, "If I do A *and* B, will you do C?" He offers Chananya a simple contract in which they agree to exchange not things but actions. Had the executioner offered to do only one action, either A or B, as so many bioethicists think he does, Chananya might have declined. Instead, the executioner surmises that he would need to offer both A and B in order to secure Chananya's participation in the plan and therefore get what he himself wants.

And get his partner's assent he does. Not once but twice the executioner inquires whether Chananya agrees to the conditions of their contract. This duplication is suggestive, especially for care at the end of someone's life. It is one thing to propose a particular treatment plan and get a patient's or family's assent to it. It is quite another to secure a second round of endorsement. Duplicating positive support suggests that the proposer's and patient's interests truly align. A single initial nod might indicate only preliminary or partial consideration of a plan. A second, after having time to fully consider the ramifications of the proposal, better communicates allegiance to the plan. Though it is unclear just how much time he had to reconsider the proposal, Chananya, like the *Bat Kol* later on, gives the plan his confident imprimatur. So not only does the executioner speak up and thus appear all the more human, he also engages with his dying charge several times with utmost respect. He offers a multi-prong treatment plan and patiently labors to elicit clear consent from his counterpart. Though his motivations are complex, perhaps something about his relationship with his charge can serve as a model for thinking through contemporary scenarios of euthanasia.

Reflexive Relations—Patient

Chananya does not die instantaneously in the story. His is a lengthy demise, beginning even before our particular *sugya* when the Roman court found him guilty of a capital crime. And his attitude about his ultimate end vacillates throughout the narrative.

Upon emerging from the court he immediately justifies his fate by invoking scripture. Such piety, however, comes before the actual pain and suffering he will no doubt endure. So, too, do patients frequently pray before surgery. Once in the heat of the moment, however, scripture slips from his lips though the Torah burns onto his heart. To his distraught daughter he offers words of comfort, as if to distract her (and himself perhaps) from his current condition and refocus on larger issues. His students vie for his attention, plying him with a question about what he sees—a query that seems painfully impersonal. For unclear reasons they encourage him to asphyxiate himself, a suggestion that Chananya forcibly resists, as if to reassert his autonomy and sense of self despite and because of the difference between his experience within the fire and theirs beyond it. Yet his own principle requires him to endure his sufferings even longer. Such certainty and rigid independence melt, however, once the executioner pipes up with his alternative to the status quo. Now Chananya seemingly ascertains the gravity of his situation: his end is nigh. But not nigh enough. So he agrees to the plan that in many ways contravenes the very principle he articulated to his students. Even if his assent appears like a contradiction to his earlier position, it functions consistently with his current state of affairs. It could well be argued that such momentary internal–external consistency constitutes personal integrity. But since Chananya was incapable himself of enacting the executioner's role in the executioner's plan, his assent to the plan serves as a de facto deputization of the executioner to do his will. With hardly any words he assigns the executioner as a *shaliach*, his emissary bidden to do what he cannot. Since the Talmudic rabbis held it impermissible to send emissaries on transgressive errands—a teaching Chananya probably knew well, we can surmise that Chananya's mission for the executioner is not inherently wrong or bad.[102]

This closer look at Chananya's progression discovers a man undergoing extraordinary pressures from without and within. Along the way he alters his attitude about his inexorable end. He shifts from praising its proclamation to looking anywhere else but at it, from resisting it to succumbing to its inevitability, even welcoming its hastened arrival. He obviously experiences ambivalence, strongly held apposite convictions. The text may not explicitly describe his emotions as his flesh burns from his bones, but we readers of the text are free to imagine his acceptance, denial, distractedness, rage, despair, acquiescence. Such dynamic reactions to his worsening condition and shifting interlocutors testify to his emotional vitality. Chananya lives vibrantly precisely as he dies to die.

The impulse among most bioethicists to see in him a character so flat as to be unidimensional or achronic (outside time) snuffs out this emotional dynamism that is also readily apparent in many if not most dying patients today. And the urge of bioethicists to read Chananya synchronically, that is, conflated into or confined to a singular moment or slice of this story, denies to him the diachrony the *sugya* takes pains to record and that he surely lived through if he did live. Their desire for him to be consistent through time meets with difficulty the text's insistence that he is inconsistent in time. If the story does not say what they want it to say, the bioethicists report it as if it did, leaving details of his life and dying on the cutting board. In so doing they rupture his story for their agendas; they lose track of his dying, his death, his life, his story. The temporal nature of his biography—even, and especially, the very last moments of it—constitutes the narrative unity of his life.

Appreciating that unity in its fullest is the least we modern readers can do.

Reflexive Relations—Intervener

The executioner's very appearance in the story is curious. The *sugya* could have ended with Chananya's *mutav* statement, his principled lesson to his students. But no. The story goes on to include the executioner who cannot remain silent while going about the business of killing yet another rabbi. In the very act of speaking up, he expresses anxiety about his role in bringing about the death of this individual. While his proposal has been well studied, why he concocted this proposal at all has not.

His goal—so he says—is to gain access to the World to Come. He is not the only gentile to seek refuge in Jewish theology, as discussed in Chapter 4. But why does this particular, anonymous executioner seek this specific other-worldly reward? Might it be that he can no longer stomach the cruelties Rome reigns down upon its denizens, citizens, and aliens alike—and that he wants to get out of it? Answers to these questions remain obscure. Yet they must be addressed if we are to take the executioner, and his words, seriously: he is motivated to intervene in Chananya's demise for reasons that have little to nothing to do with Chananya himself.

If this is true, what might it say about contemporary interveners? Just as we suggest looking at the characters of this story as fully embodied humans complete with unique histories and complex motivations, the executioner's goals also merit consideration especially when developing norms based on his behaviors. Hence it stands to reason that modern interveners who may emulate his actions (e.g., withdrawing life-sustaining treatments, and applying lethal treatments)—as these are endorsed by the patient and by heaven itself—may function with such complex, even theological, goals in mind that are mostly unrelated to proximal patients.

Of course some care providers may rankle at this analogy of being compared to an executioner. This resistance, however, echoes the executioner's own anxiety about and distaste for his role in the regime's killing machinery. He perhaps wants to get out of it. He perhaps wants something else, but in order to get there, he needs to do one last thing: hasten this particular person's death. Today's clinicians and care providers may discover within themselves similar anxieties about intervening in another's death. Like the executioner, they may be motivated by goals unrelated to the patient at hand, but they nonetheless cannot forsake their responsibility to attend carefully to that individual.

Recognizing the executioner as a human being endowed with personal motivations that transcend this particular moment enables the modern bioethicist to view today's care providers as persons as well. All interveners are persons, not automatons. Their aspirations and goals transcend the peculiarities of any particular moment. However much we may want them to act from specific kinds of motivations, this story reflects the reality that persons behave for reasons we may not fully understand or endorse. Acknowledging this fact, however, would bring care providers more fully into bioethical deliberation than would denying it.

Narrating a Death

The story of Chananya's death as it is told in the Babylonian Talmud begins before the *sugya* bioethicists commonly cite and it extends beyond the conclusion they typically envision. It is open-ended on both sides. Each person's life is similarly open-ended insofar as each emerges onto the world scene and into a communal and familial story begun long beforehand, and each steps off the world stage leaving others to carry on their memories and concerns. Narratives as much as lived lives are boundless, and they overflow with real characters changing in and through time.

For all these reasons, it behooves contemporary bioethicists to read the story of Chananya's last moments in the fullest ways possible. Insofar as bioethicists aim to influence real human beings living and dying today, why not read this story, as it is recorded, *as if* it pertains to real people as well?

This proposal may seem rather naïve, however. For what if Chananya did not exist? What if he is only a figment of the Talmudic editor's imagination—a convenient rhetorical trope trotted out when so desired? Especially since there are so many competing versions in the Judaic textual tradition of how and why Chananya dies, perhaps a more sophisticated understanding of this story is that this particular story interweaves real and fantastical elements. Ricoeur speaks of the need to appreciate stories, biographical and fictional alike, as concoctions of truths and untruths:

> As for the notion of the narrative unity of a life, it must be seen as an unstable mixture of fabulation and actual experience. It is precisely because of the elusive character of real life that we need the help of fiction to organize life

retrospectively, after the fact, prepared to take as provisional and open to revision any figure of emplotment borrowed from fiction or from history.[103]

It could very well be that the famous Talmudic story cited by bioethicists, assumed for its reality, and revered for its supposed normative stature, is itself an "unstable mixture" of fact and fiction.

But does acknowledging this possibility require changing the way one reads the story? In which ways can and should bioethicists appreciate this story? Or should only certain sections of it be considered real (enough) to merit consideration for norm generation? Who decides which elements those are, and by what criteria are those allocations made? These concerns raise serious hermeneutical, philosophical and political issues that can be bypassed by the simple solution to abandon the story altogether because it might—and probably does—contain the unreal and the untrue. But if we took this route, what would Jewish bioethics look like, and how would it be distinctly persuasive, especially in this modern era in which rhetoric and not coercion rules? We consider such questions in the next chapter.

CHAPTER 6

SALVAGING STORIES IN AND FOR JEWISH BIOETHICS

Under conditions of adversity, individuals often feel a pressing need to re-examine and re-fashion their personal narratives in an attempt to maintain a sense of identity.[1]

INTRODUCTION

That a single narrative can inspire various and diverging interpretations is now obvious. Stories are by their nature ambiguous; they lend themselves to open-ended engagement. The story of Chananya's end is no exception: its complexity permits, even demands, multiple perspectives and interpretations. Its ambiguity is a generous resource, for certain. The ambivalence it expresses, however, dogs those modern Jewish scholars who read this story in search of a single perspective, definitive interpretation, or clear norm, which, for the better and for the worse, is what most contemporaries attempt to do.

This pursuit of simplicity undermines the richness of the story. Certainly, the emotional range expressed in the *Avodah Zarah* narrative receives scant attention as do many important details and characters, and the other versions of his demise barely get mentioned. It is as if the vast Judaic textual tradition is comprised of only this singular source, and within this source only a few features of this story are deemed worthy of scholarly attention. Such winnowing down of the Jewish library generally and this story itself hampers efforts to offer rich and enriching Jewish scholarship, no matter if it is descriptive (as in the theo-political camp's focus on martyrdom) or prescriptive (as in the bioethical camp's focus on euthanasia). Narrowness breeds shallowness.

This might lead some to conclude that stories should be abandoned. Narratives only distract, little ditties that they are, from the important tasks of the day. Since they are opaque and permit contradictory interpretations, their normative value is nil. Negligible, too, is their descriptive worth, because if we take all the versions of Chananya's death into consideration, it is truly unclear what happened to him. This story—as perhaps for all stories—is but children's folklore, a fools' tale. Vapid, flat, miniscule in importance, we should just let such stories disappear.

Such a move would be foolish. For it would deny the fact that the Jewish textual tradition overflows with stories. If we excised them all, the library would be quite skimpy indeed. Furthermore, Judaism's rules and strictures would seem all the more questionable because, as Cover would argue, there would be no overarching narrative universe connecting them and providing them communal meaning.[2] If laws can be likened to bones (e.g., *lex talionis* and laws with teeth), the removal of fleshy stories would reveal a skeleton so weak for lack of cohesion, its collapse would be imminent. Stories put more than flesh on legal bones; they provide ligatures connecting laws and values, and in the process coordinating them. Stories animate law. They help explain how and why certain laws came into being, and, perhaps even more critically, they intimate where such laws can take the civic body. Stories embody norms' provenance and *telos*. Stories render norms sensible.

Some might retort that narratives are superfluous when it comes to Jewish legal decision-making as such. Whatever legal skeleton the tradition may have, it suffices to meet each and every challenge humankind could imagine. This attitude has champions among some Jewish scholars, with J. David Bleich being the most prominent legal positivist. For them the best way to attend to morally perplexing circumstances is to plumb the legal tradition and discover therein appropriate answers. They use a variety of sophisticated legal hermeneutics and analogies to uncover the tradition's response to a novel problem. Law for them is the only source from which norms derive.

Or so they hold. This project has a bone to pick with this formalist claim of legal purity. Whether conscious of it or not, all Jewish ethicists utilize stories in their work. Like all Jewish bioethicists, J. David Bleich reads, interprets and incorporates stories in his bioethical tracts. The prevalence of stories varies across the field of Jewish ethics, to be sure;

the subfield of bioethics exemplifies this clearly. And how scholars read narratives and the ways in which they inscribe them with normative influence certainly differs. Thus the concerns that began this project remain pressing: how do modern Jewish bioethicists read stories, and what might this mean—to the stories, to bioethics, and to the intended audiences? If the ways bioethicists read and utilize stories are as unhealthy as has been shown in previous chapters, are there ways to improve bioethical readings of stories?

This last chapter ascends through these and other questions in three large bounds, each with three smaller steps pertaining to texts, reading, and writing. The first leap is over some remaining burning problems regarding this particular story of Chananya's fiery end. Some residual yet significant textual challenges demand attention, as does the very practice of reading this text for its medical insights. No less worrisome are the extant writing efforts that flatten this otherwise dynamic story into a timeless norm. Leaving these issues unattended would be dangerous, I contend, for the text and textual tradition generally, for bioethicists and their ilk who like to read—and read into—stories, and for audiences who seek guidance on especially complicated moral conundrums like end-of-life care.

The second large bound offers some salves for these simmering issues by turning to modern narrative theory, social psychology, and philosophical bioethics. We chimney through ideas about appreciating texts as they are, reading stories, and integrating narratives and norms. These ideas contribute to the last substantial move in which we return to Jewish bioethical discourse proper. Specifically, we explore how this complex story can be taken into consideration as a narrative per se; we offer suggestions on ways to read it that are broader than current hermeneutical habits; and finally, we situate bioethical discourse in its living context inasmuch as it is a genre connecting people in a community.

These three bounds traverse from failing structures and patterns to stronger and more promising ones. As such, this chapter gestures toward narrative repair, an effort to salvage this story—and others—in and for modern Jewish bioethics.

Burning Problems

The story of Chananya ben Teradyon's fiery death appears to be straightforward, according to most bioethicists who recapitulate and reference it. So often in their hands the story is clear and concise.

And because the story begs no questions, one can trust it. It is only reasonable to assume it can ground unequivocal norms; it establishes certainty.

A closer examination of the story betrays all this, however. Consideration of the story as it actually is in the classical source proves that bioethicists give it a short rift. They twist it, mistranslate it, overexpose certain elements at the expense of others, and cut it apart as one would dissect a corpse, disposing of presumed uninteresting and worthless pieces. Such decimation harms the story, to be sure, for much in it is thus lost. So, before we reflect on how bioethicists read (into) the story, we take Cutter's cue to revisit our story again, so as to appreciate—and complicate—the story itself.

There are three features to consider afresh here: ambivalence, ambiguity and accuracy. According to contemporary bioethical discourse, only one story of Chananya ben Teradyon's death exists and it is found in the Babylonian Talmud's tractate *Avodah Zarah* 18a. This is how the story exists in that source:

1. They said that only a few days had passed since the passing of R. Yosi ben Kisma when all the elite of Rome went to bury him and eulogize him with a great eulogy.
2. Upon returning they found R. Chananya ben Teradyon who was sitting and engaging in Torah, convening substantial gatherings, and a Torah scroll rested on his lap.
3. They brought him and wrapped him in the Torah scroll, and they placed piles of twigs around him, and lit them aflame. They brought tufts of wool soaked in water, and placed them upon his heart so that his soul would not depart quickly.
4. His daughter said to him, "Father, to see you thus!" He replied to her, "Were I to be burned alone it would be a difficult matter for me. Now as it is me who burns and the Torah scroll is with me, the one who will address the affront to the Torah scroll will also address the affront to me."
5. His students said to him, "Teacher, what do you see?" He replied to them, "The parchment burns but the letters soar." "Also you can open your mouth and the fire will enter you." "It is better that the one who takes it is the one who gave it. One should not injure oneself."
6. The executioner said to him, "Teacher, if I increase the flames and remove the tufts of wool from your heart, will you bring me to the World to Come?" He said to him, "Yes." "Swear to me." He swore to him. Thereupon he immediately increased the flames

and removed the tufts of wool from his heart. His soul quickly departed. He even jumped and fell into the fire.

7. A *Bat Kol* [a heavenly voice] came and said, "R. Chananya ben Teradyon and the executioner have been assigned to the World to Come."
8. Rabbi [Judah HaNasi] cried and said, "Some acquire his [eternal] world in one moment, and some acquire his [eternal] world in many years."

Without rehearsing the myriad ways bioethicists hone in on Chananya's *mutav* statement (paragraph 5 above) or the executioner's proposal (6) or Chananya's assent to that plan (also in 6), certainly there can be no doubt that this story is more complicated than how bioethicists usually re-present it. Recall that before this particular story begins, Chananya and his family emerged from the court in which their punishments were meted out and they extoled the justness of their fates. Such trust in human (and divine) justice—even in Roman justice—is therefore this story's context. How might this context frame these ensuing events? Though one could argue that Chananya's certainty never falters throughout the story, and this argument could pass muster, it does not do justice to the ambivalence that he nonetheless articulates within the confines of this *sugya*, within the story as it is encapsulated here.

Notice, for example, Chananya's silence as he is wrapped within his Torah scroll and placed atop the pyre (3). Where is his proud certainty now? Could his refusal to make such comments now be a gesture of protest? Demurral is more than plausible in light of the next instant when he offers words of comfort and consolation to his distressed child (4). Remember that earlier she also praised the justness of her and her father's fates. Now she wails against the reality of it coming to fruition; she does not extol it. Her change of heart is painfully exposed, yet Chananya does not blame her or rebuke her for this inconsistency. Rather, his fatherly concern for her overrides his prior impulse of bravely embracing foreseen yet distant death. That kind of bravado now fades as he recognizes the fact that his demise is inexorable and imminent. What he wants to express now to his kin is solace. Ever the father, he models for her paternal tenderness appropriate for her plight, not his. Her anguish trumps his.

His students then interrupt this intimate familial moment, barging into the narrative to ask what seems to be a rather insensitive question about what he sees (5). They do not express loving concern for their enflamed teacher. Their sole interest, it appears, is selfish: they seek yet another glimpse of this man's extraordinary insight. They do not devise ways to save him by offering to extract him altogether from the flames searing his flesh; on the contrary, they encouraged him to asphyxiate himself. Surely this suggestion could be selfish since in that ancient rabbinic world leadership positions were highly desirable and hotly sought. Given this context, Chananya's response to his students seems more an expression of political self-preservation than a grand bioethical principle. The tenor of his retort sounds more like remonstration than instruction. Taken out of that political context internal to the Jewish community and situated in the even larger framework of Roman oppression, his response fails to echo the theological fervor he previously uttered. His focus on personal restraint of self-injury instead of personal or collective praise of God, suggests one should preserve oneself *despite* and *because* of the situation; God is all but irrelevant when it comes to life and death.

This turn to human agency is further enhanced when the centurion proposes a plan to Chananya (6). Though it may be unclear how much time passes between his conversation with his students and this one with his executioner, it is patently obvious that his mind has changed. His resistance to hastening his own demise has evaporated as has his flesh. That the executioner's plan is now palatable raises the question: Would Chananya so readily endorse this plan earlier—before he spoke with his daughter or his students? We can only offer conjecture on this point. But we can assume that the author of the story intended this conversation to occur precisely here, after those prior interactions; it was meant to come late in the flames. Chananya's twice-secured assent to the executioner's plan demonstrates his full conviction that this is the best thing for him to do. In a way he hires the executioner to do what he himself cannot; he will pay him according to the terms of their contract. Might this have been a theological ruse, with Chananya knowingly agreeing to do something for the executioner that he could not guarantee absolutely? Perhaps. Yet this does not reduce the importance of the fact that Chananya explicitly contradicts his prior statement to his students. To them he said self-injury is to be avoided. Now he endorses a plan that deputizes another to lethally harm him. If the former is to be abjured, all the more so lethal self-injury should be proscribed, no matter if it is done actively by oneself or passively via one's agent. His change of heart could not be more apparent.

The *Bat Kol*'s announcement that both Chananya and the executioner attain places in the World to Come (7) and Rabbi Judah HaNasi's lament about it (8), together express ambivalence regarding theodicy. They differ regarding the notion of just desserts. One articulates the appropriateness that different people merit the same reward; the other thinks this is inherently unfair. The redactor of the story apparently agrees with both perspectives, as evidenced by the fact that both are included in the story as we have it.

This cursory review of the narrative reveals ambivalence of great magnitude on a variety of issues. Theologically the story endorses both the notion that God matters a great deal in and to the vagaries of human history, as well as the rejection of this conviction. The nature and extent of human agency is hotly contested. And theodicy garners indecision as well. In sum, it is difficult to say with any semblance of certainty precisely what this story holds vis-à-vis any one of these issues.

Just as the story's internal ambivalence denies simplification, the existence of multiple stories narrating Chananya's demise points to irrefutable ambiguity. This is evident even if one were to consider the story as it is told in *Avodah Zarah* alone. The very existence of competing manuscripts—especially in regard to what Chananya says to his students—renders the now popular printed edition doubtful. Complicating this further is the fact that the Jewish textual tradition countenances at least three additional versions of Chananya's demise. As detailed in Chapter 3, these competing stories diverge from the *Avodah Zarah* version in dramatic and troubling ways. Some of these texts even predate the Talmud's final redaction, which suggests that perhaps what is recorded in *Avodah Zarah* is a corruption of earlier renditions of the story. However contestable this suggestion may be, it raises other questions regarding the veracity of any of these narratives.

And this concern about truth leads to the last and most significant issue regarding this story at the textual level: the problem of accuracy. According to Günter Stemberger in his now critically acclaimed classic, *Introduction to the Talmud and Midrash,* Talmudic narratives are inherently and seriously suspect. This is so for several reasons. The primacy of law, as evident by its prevalence in the Mishnah and other Tannaitic literature, suggests that when a law is attributed to a sage, that attribution is credible. Not so regarding narratives. This is because they are "almost always later than the halakhic materials attributed to the particular master."[3] Stemberger examines the ramifications of this fact at length:

But because the frequency of narrative material increases with the temporal distance from a particular rabbi, one must seriously expect later inventions and

> embellishments. A large part of the rabbinic narratives is useless for serious biography . . . The "biographical" narratives about the rabbis are not accurately transmitted eyewitness reports: most of them are relatively late texts intended for edification, exhortation or political ends (such as support of the patriarchate or other institutions). They are usually legendary, stereotyped narratives . . . The primary interest in individuals in these texts is only superficial; in reality their intention is above all to inculcate certain attitudes of life, the rabbinic way of life and ideal of study . . . Internally, then, the texts pursue pedagogical purposes; externally they are rabbinic group propaganda.[4]

Temporally distanced from the people they describe, Talmudic biographical narratives are more fantasy than fact. They are less descriptions of reality as it actually happened and more idealizations of what might have been. Historiographical veracity mattered little to these stories' authors and redactors. Rather, they were more concerned about preserving institutions and promoting particular values. Securing the truth of the past mattered less than shaping the present and future of Jewish life.[5] This may energize those bioethicists plumbing this story for norms, yet they must nevertheless contend with the fact that the story is not factual in total or in part. This truth leads Stemberger to conclude, "A biography [extrapolated from Talmudic materials] in the usual sense will always remain unattainable."[6] Insofar as biographical tales in the Talmud are inherently unreliable and should be considered with "serious reservations," basing norms upon them is a doubly dubious endeavor.

These burning textual issues compound complications of reading Chananya's story for bioethical purposes. Scholars of ancient Jewish medicine lament the absence of a singular compendium encapsulating extant medical knowledge. Samuel Kottek thus finds that "the biblical and Talmudic medical knowledge are usually fragmentary and incidental, and the data cannot be dated with precision."[7] Discerning medical knowledge and practice is piecemeal at best and anecdotal at worst. It requires readers of these ancient texts to decide which pieces therein are medical in nature and which might be medically relevant if they were only read in a certain way.

An initial glance at the story of Chananya's demise—especially as it is conveyed in *Avodah Zarah*—highlights this challenge. Take, for example, the fact that Chananya was punished for a crime, or at least what the Roman regime considered a crime. Analogizing this political circumstance to a medical milieu stretches the story in questionable ways, as Jonathan Cohen notes. "The accused standing before the judge faces the overwhelming force of the state and is no longer in control of his destiny. On the other hand, the patient meeting the doctor presumably can reject medical advice, refuse treatment, and sometimes insist on the continuation of therapy against medical advice."[8] Chananya could not wish away the state's capability to kill him; he could only adjust how and when that demise ultimately would come about. He could no more reject the state than he could conduct his own death against the state's plans. No patient was he.

Importing this explicitly political story into medical deliberation obliges readers (bioethicists, that is) to justify why it should be read (medically) at all. James Childress insists that readers need to identify those features of the source story (Chananya's demise) that are common with the target situation (euthanasia) if they want to claim that the former has any normative claims on the latter.[9] It is unsurprising that bioethicists pick and choose certain features they consider readily corollary to medical circumstances—at the expense of reading the story in its totality. However understandable this selective reading practice may be, it still leaves much to be desired. For example, if the exchange between Chananya and the executioner is a reasonable feature commendable to deliberations on euthanasia, it would be unreasonable to disconnect what is said from the people who say it. Thus, what can be made of the fact that the executioner *is* an executioner? Does this mean modern bioethicists consider contemporary physicians—people professionally trained to sustain and improve people's lives—akin to executioners—people professionally trained to end and critically compromise people's lives? Some bioethicists probably do want to make this analogy, especially if they consider euthanasia condemnable: any physician participating in that kind of intervention should be considered a kind of modern-day executioner, albeit in a medical and not political setting. Yet this further aggravates the analogy, since the source story is about a state actively killing a victim whereas euthanasia usually pertains to people—and not a government as such—struggling with how best to care for a dying patient. Few bioethicists overcome this analogical challenge and fewer still acknowledge it in the first place. All this suggests that the larger question of this story's relevance to medicine remains problematic.

Such doubts about the relevance of this story or its features to medical concerns notwithstanding, bioethicists regularly invoke it in their deliberations on euthanasia. As has already been noted above and previously detailed in Chapter 5, when bioethicists write about this story they frequently extract time from it by fixating on one or another feature of the story as if that feature alone is the totality of the narrative. Few take the effort to treat the narrative in its fullest presentation (as it is found in *Avodah Zarah*), and only one or two wrestle with the fact that the story most certainly tells a tale that traverses time. The tendency to flatten the temporal dimension denudes the story of its narrative qualities. If only it could be a singular moment or teaching, it could be law, a norm that transcends time, stitching one historical moment's norms into the future.

Writing the story as if it has no internal temporality is especially difficult due to its complex narrative structure. In his study of Talmudic legal narratives, Barry Wimpfheimer finds that, "The more dialogical such a narrative is—framing the normative in the context of conflicting semiotic characterizations—the more palpable the sense that the narrative resists the authoritarian single consciousness of a uniform legal discourse."[10] To be sure, this story about Chananya's death is not a legal narrative per se. That said, its dialogical structure is readily apparent, giving voice to diverse perspectives. For example, the *Avodah Zarah* version has no less than six characters explicitly speaking, seven if we include the latent author describing the scene as it initially unfolds. With so many voices vying for our attention, the story resists condensation into a "single consciousness" of any sort—narrative or normative. Bioethicists ignore the polyphonic nature of the narrative and instead hear and convey only one voice—the voice with which they agree. That is, they hear in the narrative their own voice and this is the one they write about.

Another curiosity regarding the bioethical use of this story regards the presumed role of authority. It seems most bioethicists assume the story is medically relevant; if it were not, they would probably not cite it at all. If we take as given its purported relevance, and granting that one or another feature is salient to euthanasia in particular, the next question is who is authorized to do anything about intervening in another's demise. To answer this, we must differentiate between the active curtailment of someone's life and the shortening of someone's dying process. Fred Rosner, medical doctor that he is, frames the issue thus: he permits withdrawing life-sustaining instruments if and only if "one is certain that in doing so one is shortening the act of dying and not interrupting life. Yet who can make the fine distinction between prolonging life and prolonging the act of dying? The former comes within the physician's reference, the latter does not."[11] On the one hand physicians are qualified and are the rightful authorities to prolong life: they must do their utmost to keep a patient alive or animated. On the other hand, physicians are denied the authority to prolong someone's dying process. If they don't have such authority, who does? Though Rosner resists explicitly answering this question, J. David Bleich is not so shy. Bleich insists that rabbis are the sole rightful authorities to decide such matters because they are more Judaicly knowledgeable than physicians, and since they are more distanced from a particular patient, they perforce are more dispassionate when rendering such decisions.[12] Deference to the rabbi is a must for such scholars, even here when the matter at hand revolves around a guttering life a particular rabbi may never have met. If it is a rabbi elsewhere who has the sole authority to make normative decisions about this particular patient's dying, it is as if, in writing this way, these and other bioethicists marginalize if not erase altogether the real physicians, families, and even the dying patients themselves from consideration when it comes to making rules regarding euthanasia. Certainly this jurisprudential assumption protects the supposed purity of legal formalism, but it does so at the expense of silencing the real people it strives to influence. Many of these real people may find this so troubling that they reject wholesale what such distant rabbis have to say about euthanasia generally or about a specific case in particular. Instead of reactionarily holding that rabbis should have nothing to say about norms regarding euthanasia or about a particular case, perhaps they *and* the real people typically involved in such cases should all be authorized to speak and co-determine what can and should be done.

SALVES FROM NARRATIVE THEORY

Despite these burning issues, not all is lost to bioethical discourse when it comes to Chananya's dying and death. Possible salves already exist in a variety of different fields that contend with narratives; the challenge and opportunity here is to integrate them in modern Jewish bioethical discourse.

Stories often include a variety of discourses. *Reported* discourses, for example, are those in which the narration relates events as they happened. *Transposed* discourses summarize what characters say. And *narratized* discourses quote characters. Teasing apart these kinds of discourses improves our understanding of a narrative's structure. Were we to apply them to the *Avodah Zarah* version of Chananya's demise, we can see all three kinds at play. The beginning of the story ("They said . . . ") reveals that what ensues is an oral tradition: at the macro-level it is a transposed story. Embedded within the whole, however, other kinds of discourses emerge. The first several moments (paragraphs 1–3 above) report what happened, the rest (4–8) include narratized discourse, and hidden within one (6) is some transposed discourse ("He swore to him"). In many of these moments (3–6) actual conversations occur between characters, regardless of whether we consider them too brief to merit the epithet "conversation." Other spoken words (7–8) seem to be articulated to no one in particular but to all and sundry.

Insofar as norms apply to a population generally, perhaps we should only look toward those elements of the story that are spoken not to individuals but to the public writ large. This obviously would dissatisfy many bioethicists, as it would mean skipping over most of the story and leaning only or heavily on the sayings of the *Bat Kol* and Judah HaNasi. So what about leaning only on direct quotes, the specific words of characters? Though attractive for its concreteness, this approach runs up against the fact that in many instances what is said is contestable. This is especially true in regard to Chananya's response to his students' suggestion to asphyxiate himself. It is further complicated if we consider what he says in the other competing versions of the story. And for those who rely upon Chananya's response to the executioner, how trustworthy is his transposed discourse? What precisely did Chananya swear? We do not know because the story's author hides Chananya's actual words. This muffling had to be purposeful. What might this erasure or suppression mean for bioethics?

These kinds of discourses also exist in the other versions of Chananya's demise, though their relative prevalence differs (for a refresher of these, see Chapter 3). *Sifre,* for example, opens with reportage, moves into several moments of narratized discourse between an unspecified "they" and Chananya's family members, pauses to hear Rabbi Judah HaNasi praise them to no one in particular, and concludes with a heated narratized conversation between a philosopher and a governor. The *Kallah* text, in which only Chananya, the good executioner, and the *Bat Kol* appear, uses only narratized and reportage styles of discourse but adds one more: *internal* discourse. Here we are able to listen in on the executioner's internal deliberations. The *Semachot* version also incorporates reportage and narratized discourse to tell the story of Chananya and his daughter conversing about the theological justness of his fate.

If transposed discourse is the most suspect kind inasmuch as a story's author summarizes what a character says, it would seem that these other versions of Chananya's demise are less susceptible to human bias than *Avodah Zarah,* specifically in regard to Chananya's response to the executioner's plan. On the other hand, to the degree that rendering decisions about end-of-life care necessarily involves multiple people and it is hoped that those people interact and communicate with each other, the text that offers the richest array of conversations across various stake holders is indeed *Avodah Zarah*. If that text is deemed relevant to medical circumstances—itself a conclusion still awaiting non-question-begging foundations—then it makes modest sense to look carefully at that text's conversations to see who speaks to whom, when, about what, and what ultimately transpires.

Taking a step back a bit from the narrative's elements to its broader sweep, it should not be lost on us that Chananya is under extraordinary distress. Mike Bury has observed in illness narratives that people in extreme circumstances often adjust their personal stories so as to make sense—to themselves and to others—of what is happening to them.[13] Chananya appears to fit this pattern since he changes his mind about the palatability of someone intervening in his expiration. In this way Chananya leaps off the page of *Avodah Zarah* as a real, recognizable human precisely because he responds to and through his suffering. The men called Chananya in the other versions seem more staid and stubborn; they do not transform as much as this one.

Reading this story for its realism is tricky, however. In the view of philosopher and bioethicist John D. Arras, stories that are real are really difficult to characterize or compartmentalize. Their inherent complexity defies easy theorizing and clear norm extraction. Hypothetical stories are different, however,

> Because hypothetical cases, so beloved of academic philosophers, tend to be theory-driven; that is, they are usually designed to advance some explicitly theoretical point. Real cases, on the other hand, are more likely to display the sort of moral complexity and untidiness that demand the (non-deductive) weighing and balancing of competing moral considerations and the casuistical virtues of discernment and practical judgment (*phronesis*).[14]

Though his is an argument commending the re-emergence of casuistry in modern bioethical discourse generally, Arras notes that the two dominant kinds of cases—hypothetical and real—serve different purposes.[15] In his view the latter are better for bioethics insofar as this field is meant to grapple with the messy and fleshy issues of lived life. Details and peculiarities—these make cases interesting and biomedically relevant. True though this may be about real cases, it behooves bioethicists not to rely upon only a singular real case when rendering a norm. It is not sufficient to illustrate a point by a case. Rather, several cases should be consulted so as to "see just how reasoning by paradigm and analogy takes place and the so-called 'principles of bioethics' are actually shaped by their effective meaning by the details of successive cases."[16] Arras thus lobs a two-fold critique toward bioethical discourse. Those who are inclined to read real cases for principles and norms perforce strip them of their peculiarities and actual people, voiding them of much of their reality; that is, they read such cases as they are not. And those who prefer to keep the complexities of real cases front and center, they should not contend themselves with consulting only one case or arbitrarily elevating one as paradigmatic against which all others must compare.

These concerns encourage Jewish bioethicists to take the "moral complexity and untidiness" of Chananya's dying both seriously and not exclusively. His morally messy demise merits a close treatment, for sure, yet it need not be considered the primary paradigm for deliberation on euthanasia. The impulse to read it as the *exempla par excellence* of rabbinic anxiety or certainty about intervening in someone's dying should be resisted. James Childress echoes this concern when he says, "archetypal narratives may obscure individual narratives and thus seriously distort what a particular patient is saying."[17] Conversely, Arras's arguments challenge Jewish bioethicists to wrestle with the fact that Talmudic narratives are most likely not eye-witness accounts. Their predominantly legendary nature suggests that their authors probably had some intentions, perhaps theoretical in nature, in constructing them the ways they did. If this is true, bioethicists need to consider the totality of the narrative to discern what those "theoretical points" might be. As demonstrated throughout this book, piecemeal reading does not suffice.

Other scholars also warn against reading messy narratives with clear normative intentions. The practice of reading narratives "as if they were legal statutes," says Wimpfheimer, "seeks and consistently finds the rule of law in a story that may or may not contain it."[18] Classic Jewish narratives "provide only a very general moral direction," Newman contends, "not a course of action to follow."[19] And Childress stresses the difficulty of determining "exactly where to put the interpretive weight" when hearkening to an ambivalent patient's narrative—whether ancient or modern.[20] Unethical listening, in Childress's view, is a kind of listening or reading that determines beforehand what will matter in a narrative; ethical listening, by contrast, is not devoid of interpretation altogether but it is "as careful as possible to hear what this particular person is saying, what his or her story is, without acting under a larger story that putatively illuminates the patient's particular story."[21]

Given that humans cannot but read stories through one kind of hermeneutical lens or another, it is not surprising that those desiring to see a norm in a narrative miraculously succeed nearly every time. Recent research in moral reasoning sheds light on this phenomenon. Social psychologist Ziva Kunda observes that "people rely on cognitive processes and representations to arrive at their desired conclusions, but motivation plays a role in determining which of these will be used on a given occasion."[22] There are two major types of cognitive motivation or goals: "[1] accuracy goals lead to the use of those beliefs and strategies that are considered most appropriate whereas [2] directional goals lead to the use of those that are considered most likely to yield the desired conclusion."[23] Studies show that when people are motivated to be accurate, they expend more cognitive effort considering the breadth and depth of available data, taking time to contemplate *what is*. People motivated to reach a particular conclusion, by contrast, "attempt to be rational and to construct a justification of their desired conclusion that would persuade a dispassionate observer."[24] These goal-oriented thinkers consider only those data that comport with and reinforce their predetermined goals; they contemplate primarily *what they want to see*.[25]

As this is what we do in our daily lives, we can reasonably assume we do something similar when we read. Some of us read for accuracy; some read pursuing goals. The latter find in a text what furthers our goals—those few pieces that reinforce our assumptions, attitudes, convictions, conclusions; bluntly: we quickly find ourselves in someone else's story. The former, however, plod through a text, immersing in its totality and complexity. Whereas accuracy readers carefully attend to the breadth and depth of a story, goal seekers cherry pick those pieces that appeal. "It is always easier to accept an analogy," Heidi Mertes and Guido Pennings argue, "when it is consistent with acquired beliefs and leads to a desired outcome."[26] Though moderately easier than accuracy-reading, such conclusion seekers are burdened to justify the reason that they read a text just this way; they need to explain why their goals take precedence over the author's.

These observations by social psychologists should exercise Jewish bioethicists. If they are going to read a particular narrative, say, about Chananya's death, with a particular goal in mind, say, prohibiting any and all acts of euthanasia—they need to be cognizant that they are indeed doing just this. Furthermore, they need to make this choice explicit and explain why. Without such disclosure, bioethicists misconstrue their goal-oriented reading for accuracy. They obscure what is there for what they want to be there. They impose their orientation upon the text—as well as upon their own readers. In a Chekhovian way, theirs is a personal morality in search of traditional endorsement and public agreement. Theirs is a cart preceding the proverbial—and textual—horse. Several authors, Len Sharzer and Jonathan Cohen included, lament that many Jewish bioethicists do just this.[27]

Another significant challenge for writing Jewish bioethics in light of this or any other classic story regards the practice of analogizing. Mertes and Pennings identify three kinds of analogies in normative bioethical discourse. The most common are similar analogies, in which commonalities between a classic source and a targeted contemporary circumstance are highlighted; dissimilarities are dismissed as morally irrelevant. Dissimilar undermining analogies, by contrast, highlight discrepancies between the source and target, but these analogies do not offer guidance for evaluating the target morally.[28] The third kind, dissimilar reinforcing analogies, do, however. These focus "on the dissimilarity between source and target but nevertheless aim at a similar moral judgment in source and target, as the dissimilarity indicates that the moral judgment made in the source situation is *even more* appropriate in the target situation."[29] These last kind of analogies have the most potency for the reason that they "pull the target situation to either pole of the continuum between morally acceptable and morally unacceptable situations, thus defending a stronger conclusion."[30]

The implication of this for Jewish bioethical writing is two-fold. On the one hand it is possible to critique the invocation of this Chananya story because analogies are, by definition, incomplete and not exhaustive. The imperfect match between a source and a target perforce requires selecting certain elements of each for comparison, interpretation, and conversation. This very selection is inherently suspect, for it could be that the "wrong set of elements has been isolated because not all the morally relevant elements were taken into consideration and/or because the included elements were not morally relevant."[31] Hence, the criticisms outlined in Chapter 5 that queried the moral relevance of hinging norms only on Chananya's response to his students, or on his response to the executioner. There may be other, more morally relevant, features to consider—like his conversation with his daughter—which perhaps should receive greater moral weight since family members are more likely to make or contribute to end-of-life decisions of a patient than are professional peers or even students.

On the other hand, we can critique a particular analogy by the way selected features are interpreted. This fuels the vast majority of contemporary Jewish bioethical discourse, to be sure: authors disagree about how best to interpret the few popular features without questioning the moral relevance of those features. I propose a slightly more wholistic approach that takes into consideration the manuscript variations of the story. For example, if Chananya's response to his students really can or should remain a morally salient feature in this debate, then at least bioethicists should contend with the manuscripts that have him say something dramatically different—as noted in Chapter 3. These alternative statements bespeak a different conviction than that which is recorded in the printed edition, the interpretation of which promises to enrich contemporary discourse in ways that reinterpreting the standard text (yet again) might not.

Complicating the use of ancient stories as pivotal analogies for contemporary bioethical argumentation is residue, that is, what is left by the wayside. As has just been noted above, this certainly refers to concrete elements and features of the narrative that are suppressed while other elements are highlighted and examined. It also refers to broader themes that are intimated but not explicitly treated in the narrative. Take the issue of autonomy as an example. Leonard Kravitz and Peter Knobel independently cherish the notion that Chananya responds to the executioner out of his own volition.[32] In their views, Chananya acts with some semblance of autonomy, of self-generated decision-making regarding himself. It is understandable that these authors examine this theme, just as it is unsurprising that Jakobovits did not half a century earlier. This is because autonomy only became a significant concern in bioethical discourse after Jay Katz published his now-classic *The Silent World of Doctor and Patient* in 1984 in which he took to task the healthcare industry's practice both in the clinic and in research experiments of hiding things from patients and subjects and thus limiting or undermining their autonomy. Though we should not fault Jewish bioethical writers for not attending to certain themes when they invoke and wrestle with a particular story, we nonetheless can hold all authors to account for not reading the story *as it is* in the text they consult—in its context and totality. For what is left by the wayside, this textual residue, Wimpfheimer rightfully notes, resists being flattened into any particular norm and, even more, it bespeaks a cultural world that also can and perhaps should make normative claims on today's audiences.[33]

Burning Alternatives

In this final section we return to Jewish bioethics proper to uncover more indigenous suggestions to improve the way stories are appreciated, read, and written about.

At base, bioethical conundrums are messy both in detail and in context. They defy simplification and clear compartmentalization. Speaking to and reflecting this reality is best achieved through stories that themselves are messy. For this reason real narratives hold more promise for making meaningful arguments than do hypothetical ones. Their untidiness should be celebrated rather than deplored. This is because the ambiguity inherent to narratives is, in Sharzer's view, their "most valuable characteristic, that and the fact that stories by their very nature are, like the problems we face in the bioethical realm, context-sensitive."[34] Because life itself is so complex and often riddled with ambiguity, the stories we tell about it and bring to bear on it should reflect and speak to that ambiguity.

Taking this comment seriously in light of the narrative of Chananya's demise, Jewish bioethicists would do well to embrace Chananya's explicit ambivalence. His fluctuating attitude about intervention in his dying speaks to and of the impulse of people in extreme pain to change their stories, as noted by Mike Bury at this chapter's outset. Denying this realistic aspect of Chananya's dying moments does violence to the story; it makes it more unreal, more fantastical and hypothetical than it already is. It also suppresses the fact that time transpires in the narrative. If at time T_1 Chananya felt one way about hastening his demise (e.g., his *mutav* statement to his students), it cannot be denied that his conviction was different at time T_2 (e.g., his contract with the executioner).[35] Without embellishing or quelling the story in any way, it is unarguable that Chananya changed his mind based on internal experiences of his dying process that happened in and through time.

Time, experience, and change of heart are also undeniably evident in another popular story about end-of-life decisions. This story, detailed in Chapter 2, concerns Rabbi Judah HaNasi's handmaid.[36] At first she prays in concert with his disciples that her master should remain among the living despite his chronic and extremely painful gastrointestinal ailment. But upon witnessing her master's extraordinary discomfort from his condition, she changes her tune and begins to pray for the immortals to take this mortal away from earth. Though it is unclear whether the disciples saw their' master's pain as clearly as the handmaid, they nonetheless do not change their prayers. They do not learn, as Sharzer observes, from their "patient"; they remain obstinate that theirs is the right and only position vis-à-vis this dying man.[37] The handmaid, by contrast, changes her mind based on her observations of the man for whom she cared. Change of opinion based on experience is both understandable and realistic. Instead of suppressing ambivalence and anxiety as many bioethicists tend to do, acknowledging them may empower bioethicists and their audiences alike as they struggle with and through the reality of caring for someone who is in the throes of dying.

Complementing the ambivalence integral to the story as it is are the competing stories as they exist in other classic sources. The very existence of them points to the fact that the same phenomenon—Chananya's fiery end—can be and is indeed viewed from multiple perspectives. Different characters appear across these other versions, as do their conversations and their thematic concerns. Healthy Jewish bioethical deliberations would consider all these sources, not just one. They would evaluate these Chananya chronicles, as I call them, in light of each other so as to discern the totality of rabbinic anxiety about intervening in and caring for the dying, or at least in regard to Chananya.

Reading the totality of the story is surely a first step toward healthier bioethical discourse, but it is not the only one. As Jonathan Cohen astutely observes:

> The question we would have to ask is not "What corpus of Jewish text is employed in this setting (*halakhah* or *aggadah*)?" but rather "How do we read the text?" *Aggadah* may well be the material preferred, but it might be employed as halakhic material would be in the hands of an able rabbi in conversations with Orthodox or ultra-Orthodox patients and relatives. The narrative could well be communicated in ways that suggest prescription and be stripped of its richness and openness. This would plausibly occur in settings of constraint, violence, and pain, as well as a profound thirst for healing and comfort. In short, *aggadah* would no longer be *aggadah* as our Sages understood it, and something of it would be lost.[38]

This question—how do we read a text?—is a critical question bioethicists can ill afford to ignore. When and how a particular story is invoked can infuse it with more normative force than might the recitation of a law. It therefore is necessary for bioethicists to come clean about how and why they read stories when and how they do. Cohen again: "Herein lies a great challenge for Jewish bioethicists: writers and practitioners must exhibit greater self-reflection and transparency regarding their own engagements with Jewish texts."[39]

Newman endorses this call for bioethicists to be more explicit of their hermeneutics: "It is important then that Jewish ethicists openly acknowledge that their work proceeds from some particular conception of the Jewish tradition and what gives it coherence. They should also make explicit the fact their position, whatever it is, represents only one option among many."[40] No one, not even the most educated bioethicist, can hold a monopoly on how to interpret a text, especially a narrative. Explicitly acknowledging a story's polysemy, its multiple meanings, is part and parcel to a bioethicist's intellectual integrity.[41] In addition to such humility, Newman calls for theological transparency when he requires bioethicists to clarify what, in their views, gives the textual tradition its coherence, its authority. This arches back to what was discussed in Chapter 1 in regard to theodicy. Indeed, when it comes to thinking through care at the end of life and especially when invoking this particular story regarding Chananya, bioethicists need to clarify their theories of theodicy, of how they square human suffering with supposed divine goodness. Bringing theodicy to the fore in discussions of euthanasia will only enrich and enhance bioethical treatment of this and other stories. For dying patients and their families undeniably confront these metaphysical questions as they grapple with physical circumstances they perhaps wished were otherwise.

Once theological backgrounds and hermeneutical methodologies are made explicit, bioethicists can then turn to the task of reading and elucidating how this particular story fits into and should inform contemporary decision-making regarding euthanasia. This means bioethicists need to situate narratives alongside norms and read them in light of each other. Newman argues that this strategy of shuttling—or chimneying—between lore and law, is necessary for modern bioethics.[42] The great modern theologian Abraham Joshua Heschel would concur since he thinks *halakhah* without *aggadah* is dead and *aggadah* without *halakhah* is wild.[43] "The value of *Aggadah*," Hayyim Bialik contends, "is that it issues in *Halakhah*. *Aggadah* that does not bring *Halakhah* in its train is ineffective."[44] Stanley Hauerwas also endorses the intertwining of norms and narratives:

> Even though moral principles are not sufficient in themselves for our moral existence, neither are stories sufficient if they do not generate (and sanction) formally valid moral principles. Principles without supporting stories are prone to perverse interpretation and application, but stories without principles have no adequate way of concretely specifying the actions and practices consistent with the general orientation expressed in the story.[45]

But what happens if indecision remains even after one consults all relevant laws and stories? How break a moral impasse? Gordon Tucker argues and demonstrates that *aggadah* should enhance traditional *halakhic* methodology, especially in regard to morally perplexing cases.[46] Seymour Siegel pushes this further. For him, Jewish norms like law need to be reevaluated in light of Jewish stories: "the *aggadah* should control the *halakhah,* not vice-versa."[47] Of course some bioethicists will balk at this, as would J. David Bleich. But even he cannot deny the fact that he resorts to *aggadah* to enrich, even enhance, his otherwise legalistic argumentation. Giving stories their due prominence in bioethical deliberations is therefore an indispensable and critical step.

The next step is to make explicit whether a particular source story has moral relevance to a target issue, and if so, articulate the nature of its moral weight. Elliot Dorff suggests four ways to appreciate the significance of classic Jewish sources, stories among them.[48] One is to assume that there is no relevance whatsoever. This approach may be honest in certain circumstances. It would also relieve many of trying to contort themselves to think and behave in outmoded ways. But if this were the prevailing mode regarding the totality of the Judaic textual tradition, it could lead to anarchic reasoning if not action inasmuch as law, too, would be discounted and dismissed. At the other extreme is the attitude championed by the Tannaitic sage Ben Bag Bag who said, "Turn it over and turn it over again, for everything is in it."[49] Assuming that the textual tradition is completely and exhaustively relevant means that innovation is more apparent than real; all that can be done is offer novel interpretations of extant texts. While this preserves the authority and continuity of the tradition, it requires reading texts beyond their contexts and perhaps distorting them in unreasonable ways. A third kind of relation devolves down to autonomy. Individuals should take the textual tradition into account rather than mindlessly deferring to the supposed wisdom of rabbis or other authorities. Individuals are responsible for discerning the relevance of sources to their circumstances. However liberating this attitude may be, it could nonetheless permit near-anarchy since each individual may do as she or he feels appropriate. Dorff endorses a fourth approach, depth theology, in which Jews identify foundational concepts and values applicable to the issue of concern and, collectively with rabbinic leadership, evaluate the contemporary issue in light of those precedents.

When considering the story of Chananya ben Teradyon's *crematio*, there are four possible ways to connect it with the moral morass that is euthanasia. It can be ignored altogether since we presume it—and any other ancient text, for that matter—cannot have any relevance. Its moral weight being nil, we are left free to render our decisions without it informing us one way or another. Or we can assume our decisions must comport to this text and only it (and other classic Jewish sources). This orientation may reflect and reinforce a kind of certainty, yet it fails to clarify how to handle the ambivalence inherent in the story, the ambiguity present in the textual tradition writ large, or the fact that biomedical circumstances since that time have changed. The third approach—autonomy—leaves it to the individual modern Jew to decide the moral weight of this story. Whatever attraction this approach may garner, it does little to build cohesiveness through a community, for I could argue that the story is extremely morally relevant while you could hold otherwise. I would hold that because Chananya endorses intervention in the end euthanasia is therefore palatable, and you would argue against this because you think the story only illustrates the anxiety rabbis held toward intervention, and you'd rather rely upon another story or other laws that appear more relevant to you. Depth theology corrects for the atomism of autonomy since it requires communities—lay and professionals together—to discern and decide a story's relevance to a modern conundrum. Yet balancing authority within this communal deliberation remains a procedural stumbling block, often distracting from the urgency of the situation. We may all agree that Chananya's story is relevant, but we disagree as to how it should be interpreted and by whom—and so we turn our attention to these procedural issues instead of those substantive ones regarding a dying loved one at hand. Thus William Cutter, though an ardent champion of narrative bioethics, acknowledges that the world does not wait for narrative decision-making.[50]

How might we read this or any other narrative when writing normative bioethical arguments? That there are significant problems of each of the four attitudes Dorff identifies should not mean that it is impossible or wrongheaded to utilize stories in bioethical discourse. Rather, the problem—and the solution—lies in silence, in hiding why one turns to stories (or other texts like law) at all.

Since archetypal stories often obscure subsequent ones, we should be wary, Childress advises, of invoking just any story when we write bioethical tracts. This is because any narrative perforce frames how problems will be perceived, constrains ethical reflection thereon, and indicates ways to resolve those problems.[51] As Cover notes, the selection of a story and its invocation in a normative argument is itself a norm-creating exercise.[52] Instead of invoking a story seemingly arbitrarily, bioethicists must reflect upon and make explicit why they turn to a particular narrative at all and not another. Only then would their interpretation of it gain its normative force in a non-question-begging manner.

This kind of self-criticism is crucial for bioethical discourse. At one level it reinforces the reality that even normative bioethical discourse is at most suggestive. It cannot be law with teeth. Authors of bioethical tracts may take inspiration from J. David Bleich who, despite his inclination toward legalistic argumentation, prefaces his work with the disclaimer that he renders only an analysis of ***halakhah*** and ***halakhic*** reasoning, that he does not produce ***psak halakhah,*** legal decisions per se.[53] This disclaimer allows room for others to disagree, for sure, and it also admits that bioethical arguments are just that: narratives attempting to be persuasive. This fact arches back to the opening chapters in which we discussed the two-fold nature of narratives: they are selective and interpretive. Just as narratives stitch together selected elements and in that very juxtaposition interpret them, so bioethical arguments are narratives since they too piece together selections from classic and other sources and thereby interpret them. Were bioethicists to acknowledge explicitly that they too select and interpret, they would clarify how and why stories can and should be integral to their essays in the first place.

Such admission has another benefit. It will make bioethical tracts all the more palatable to the burgeoning Jewish population, which considers itself not beholden to *halakhah*. Len Sharzer points to this when he discusses the attraction of narratives generally:

> It seems to me that narrative, *aggadah,* may be useful because it transmits values rather than rules and is open to multiple interpretations. *Aggadah* can offer guidance that may be helpful for many of today's Jews who do not feel bound by *halakhah* and who may distrust authority figures in general, but who still want to be guided by and act within a broadly Jewish framework.[54]

Bioethical authors who disclose their particular theologies and reasons for invoking this or another narrative would model the very values and practices of critical self-reflection that many modern Jews seek, and need, when confronting the challenge of making theologically difficult medical decisions. Though some modern Jews may want to absolve themselves completely from such decisions and defer to some authority figure about whether to euthanize and how, many want to be empowered to think through these issues. They want to be involved in making decisions about their lives and for their loved ones, regardless of how difficult those decisions might be.

Involving the actual people caught up in bioethical circumstances is a hallmark of relatively healthy bioethical discourse. No less a figure than Maimonides suggests something like this in his *Regimen of Health*. He implores physicians to understand themselves as promulgators not of law but of suggestions. Whereas law commands and compels, medicine recommends and warns. Ultimately, physicians must "leave the matter to the sick in the form of consultation; it is they who have the choice."[55] What Maimonides offers is a reorientation of medical decision-making since it cannot be pre-scribed like law generally. A legal stricture is promulgated by legitimate authority figures for society generally; medical advice, by contrast, is offered by professionals to those who are the de facto and thus de jure healthcare decision-makers for a specific case. These are the patients and their families. By acknowledging patients and families as the rightful agents of healthcare decision-making, Maimonides both empowers them and also delimits rabbinic authority from arrogating to itself the delusion that it has the right or place to decide for others precisely what they should (not) do in a particular situation. Rather, the most rabbis and physicians can do regarding health is to encourage patients and their families to consider all the medical and theological risks and benefits of treatments for a particular individual. This implies that general rules and laws are insufficient by themselves in the realm of health. The peculiarities of each patient should take the reins and constrain what can and should be done.[56]

This turn to the patient and family at a methodological level can be complemented at a substantive level as well. It would support the enterprise of looking to this story and other texts to speak about care at the end of life in general rather than trying to force the text to speak about euthanasia in particular. As demonstrated in Chapter 4, it is a stretch of imagination that the gruesome execution of a criminal in public is directly analogous to an otherwise private and hygienic demise of a patient. Despite this fact, it is possible to see in the story a variety of approaches and attitudes about caring for the soon-to-be-dead. Its various perspectives and inconsistent opinions reflect real messiness surrounding dying. In profound ways the *sugya* expresses fundamental anxiety about dying. Some characters cry, some seek insight, others pursue release—and the patient endures them all. Both the patient, if Chananya can rightfully be likened to one—and his purported care provider, if the executioner can rightfully be likened to one—die. Indeed, they both die in the same way: by flame, though one by choice and the other not. That both receive placement in the World to Come and that this apparent unfair distribution of divine rewards is lamented—expresses realistic anxiety about death. This story is so rich and real that it reflects the fact that humans are often as anxious about dying as they are about death itself—whether it is one's own, a loved one's, or someone under one's charge. Instead of stripping the story of this richness to assert that it offers but one principle or rule regarding one aspect or another of euthanasia, bioethicists may do themselves—the text and their audiences—a favor by appreciating it for its larger perspective, its realism that no matter how someone dies, conflicting emotions and convictions about how to care for the dying abound.

CONCLUSION

The task of salvaging stories for Jewish bioethics has yet to be completed, to be sure. Many other stories of stories in Jewish bioethics can and should be told. The construction and telling of such stories is necessary for the advancement of the field of bioethics. And it is valuable, according to Sharzer:

> Stories transmit values rather than prescribe specific courses of action and are thus well suited to dilemmas where context is of utmost importance. I would argue, in fact, that stories are the primary way we transmit values to children and an important way we communicate them to our fellow human beings.[57]

To this I would add that how we read stories, how we relate stories not of our own creation—especially when they pertain to dying and death—also convey values to ourselves, to each other, and to our children. Stories communicate values in and through their narratives, just as our retelling of them bespeaks our values, too.

Notes

Chapter 1

1. MacIntyre (1984):212.
2. Cover (1995):203. See also Cohen (2011):268.
3. This is the first book-length manuscript on narratives in bioethics. For shorter studies of this topic, see Jotkowitz (2012); Sharzer (2011); Cohen (2011); Knobel (2007); and Newman (2007); and for narrative in Jewish ethics generally, see Newman (1998) and Newman (2005).
4. MacIntyre (1984):216.
5. Hauerwas (1973):76.
6. MacIntyre (1984):221.
7. Hauerwas (1973):76; Levinas (1998).
8. MacIntyre (1984):217. For a critique of MacIntyre's theory of narrative unity, see Ricoeur (1992):157–163.
9. Walker (2012). See also Julian Barnes's Booker Prize–winning novel, *The Sense of an Ending* (New York: Alfred A. Knopf, 2011), for a beautiful story of the devastation imperfect self-narratives can wrought.
10. MacIntyre (1984):218. Hauerwas (1973):79 also integrates accountability into his version of narrative selfhood.
11. Levinas (1998):117, ad loc.
12. Goodman (2008):vii.
13. Ibid., 53.
14. Ibid., 20.
15. Ricoeur (1992):158.
16. MacIntyre (1984):219.
17. Hauerwas (1993):379.
18. Ibid.
19. Hauerwas (1973):74.
20. Cf. Gordon (1949); Brothwell and Sandison (1967); Amundsen and Ferngren (1995).
21. Deuteronomy 32:39.
22. BT *Baba Metzia* 107b.
23. *Vayikra Rabbah* 16.8. This source echoes R. Chanina's position, as well.
24. *Bereshit Rabbah* 92.1.

25. Sherwin (1987) organizes theodicies into these categories: horizontal, vertical, and eschatological. On Judaic notions of human suffering, see Steinberg (1999).
26. See Deuteronomy 28:15; M *Avot* 2.6; M *Kiddushin* 4.14; BT *Berachot* 5a.
27. BT *Shabbat* 55a.
28. See, for example, Exodus 23:25; Leviticus 26:16.
29. See BT *Baba Metzia* 85a; *Sifre Devarim* 32.5.
30. JT *Sotah* 5.5.
31. BT *Ta'anit* 21a. See also JT *Pe'ah* 8.8/37b end.
32. See Belser (2011) for a fascinating analysis of these stories.
33. JT *Berachot* 2.3/14a, commenting in part on Psalms 103:3, "[God] forgives all your sins, heals all your diseases."
34. *Moreh Nebukim* III:23.
35. BT *Bearchot* 5a.
36. BT *Kiddushin* 40b.
37. This stems either from the obligation to return a lost item (Deuteronomy 22:2) or from the command that an assailant must ensure that a victim achieves a complete recovery (Exodus 21:19–20). See also BT *Baba Kamma* 85a and 81b. See also BT *Sanhedrin* 84b.
38. BT *Sanhedrin* 73a.
39. See the amazing array of essays in Berger (1995).
40. See, for example, Preuss (1993); Jakobovits (1959); Sussman (1967); Gordon (1949); Kottek (2003).
41. See Rosner (1995):182–187 for more information on the Oath of Asaph.
42. On the disagreement on the precise date, see Bar-Sela, Hoff, and Faris (1964):4.
43. Preuss (1993).
44. From Rosner's introduction to Preuss (1993):xv.
45. Jakobovits (1959):xxxv, ad loc.
46. Steinberg (2004):29. This was originally published in the Hebrew edition of the 1998 *Encyclopedia of Jewish Medical Ethics.*
47. Aharon Lichtenstein's now classic (1975) *Does Judaism Recognize an Ethic Independent of Halakhah?* continues to spark debate. See, for example, the 2012 online symposia on this essay: http://www.theapj.com/symposium-on-aharon-lichtensteins-paper-does-jewish-tradition-an-ethic-independent-of-halakha/.
48. Newman (1998), Chapter 9; Newman (2005):140–141. Bleich (1985):542–543.
49. See, for example, Bleich (1973):107; Bleich (1976):296, 306n1.
50. *Iggeret Moshe, Choshen Mishpat,* II:74.2; Tendler (1996):57.
51. *Iggeret Moshe, Choshen Mishpat,* II:73:1, citing BT *Ketubot* 104a. Feinstein references this story again in his next *halakhah: Choshen Mishpat,* II:74:1.

52. Dorff (1998) illustrates this style of argumentation well. See also Dorff and Crane (2012).
53. *CCAR Journal.* LIX/II (Summer 2012).
54. Newman (1998):202.

Chapter 2

1. *Midrash Tanhuma, Bereshit* 1. See also Rashi at Deuteronomy 33:2 and Song of Songs 5:16; JT *Shekalim* 6/49d; JT *Sotah* 8/22d; *Devarim Rabbah* 3 (Ekev); *Shir HaShirim Rabbah* 5; *Midrash Tehillim* (Buber) 90.
2. The first commandment to humankind is found at Genesis 2:16; it refers to the permission to eat from any of the primordial trees in the Garden of Eden, except, of course, from the tree of the knowledge of good and evil. Many, like Rashi, point to Exodus 12:2, since that is the first commandment promulgated upon the community of Israel proper.
3. Though an individual law may not represent the passage of time, it nonetheless attempts to stitch time by linking present and future action to prior standards. See Gibbs (2004).
4. Walker (2012)
5. Newman (2007):185.
6. See Jonsen and Toulmin (1988).
7. Nelson (1997):*x–xii.*
8. Phelan (1996):19.
9. Newman (2007):186.
10. See Ricoeur (1995):309ff.
11. Knobel (2007):172.
12. See Childress (1997).
13. See, for example, the collection of essays in *Diné Israel* 2007, based on the 2005 conference, “The Relationship between Halakhah and Aggadah” held at Harvard Law School. See also Bialik (1917).
14. The Tannaim understood them to be distinct genres. See M *Nedarim* 4.3; *Sifre Devarim* §306.
15. JT *Chagigah* 1.8/76a; JT *Pe’ah* 2.4/10a.
16. Perhaps echoing the nineteenth-century Jewish historian Leopold Zunz, this is how Borowitz (2006):4 initially understands the term.
17. *Otsar Ha-Ge’onim, Berachot,* §271, 357; *Otsar Ha-Ge’onim, Chagigah,* §67. See discussion in Lorberbaum (2007).
18. Lifshitz (2007):17, argues that aggadah means “that which is tied up, or, in other words, hidden and mysterious”; aggadah, by contrast, is more akin to midrash insofar as it “is a form of biblical interpretation” (28). Since Borowitz’s (2006) overall project is to demonstrate the theology *hidden* within aggadah, he probably would

like Lifshitz' distinction of these terms even though he himself conflates the terms (see pp. 7–8, 26).

19. See the etymology of lore and learn in the *Oxford English Dictionary.*
20. For a history of the early scholarship that bisected halakhah and aggadah, see Lorberbaum (2007). Despite characterizing law as severe and story as lively in his famous "Halakhah and Aggadah" essay of 1917, Chayim N. Bialik nonetheless understands them to be like two faces of the same coin, like the two façades of a literary Janus.
21. See, for example, Gersonides's opinion about the importance of aggadah at *Be'ur ha-Ralbag* for Genesis 2:4–3:24.
22. Bialik (1923):4. This is especially true for Wimpfheimer (2007), whose work focuses on a subgenre of aggadah called legal narratives. Such stories are "some of the messiest bits of an already untidy work of literature" because they combine elements of both law and story (Wimpfheimer 2011:2). Their untidiness makes them especially precious for study because they "mimic the messiness of life itself" (ibid).
23. See, for example, the *responsum* on homosexuality by Gordon Tucker (2006) submitted to the Conservative Movement's Committee on Jewish Law and Standards in which he promotes an ethical decision-making strategy that for sure relies upon law and is enhanced by narrative.
24. Cover (1984).
25. Arras (1997):71.
26. Simon-Shoshan (2007). Michael Goldberg (1984, 1991) sides more with Cover: the character of a community, inclusive of its norms and laws, is best understood in light of its overarching paradigmatic story. For Jews this is the story of the exodus from slavery. See analysis in Newman (1998):194ff.
27. Newman (2007):184. See also Sharzer (2011):259.
28. Bleich makes a similar argument in the introductions to each of his manuscript collections of bioethical essays. Bleich (1983):ix; (1977):xviii; (1989):xi; (1995):ix. See also Newman (2007):184.
29. Newman (2007) argues that this is no less true for Jewish ethics, generally.
30. Preuss (1993):4.
31. Ibid.
32. Ibid., 7.
33. Ibid.
34. Ibid.
35. Preuss (1993):8, emphasis in original.
36. Ibid.
37. Ibid., 516.
38. Jakobovits (1959):xx.

39. Ibid.
40. Brody (1983, repr 2003):267.
41. For the Noahide laws, see Novak (1983). For the legal difficulty of and hubris involved in imposing Jewish norms upon a multicultural society, see Dorff (2006).
42. Jakobovits (1959):xxxvi.
43. Ibid.
44. Ibid.
45. Ibid., xxxvii.
46. This argument is put forward by R. Akiva Moses Eiger (1761–1837) in his gloss on SA *Yoreh De'ah* 336.1. He cites R. Yakov b. Moses Moelin's (Maharil, 1365–1427) *Likutim,* as support for his argument.
47. This is found in Jakobovits (1959):xxxviii.
48. Jakobovits (1959):xliii.
49. Ibid., 122.
50. See Newman (1998), Chapter 9.
51. See Cutter (2011) for diverse demonstrations of this kind of reading/writing.
52. Rosner (1979). Page numbers refer to the 1979 reprinting.
53. Compare with BT *Niddah* 27a where discussion is had about pregnancies of twins wherein one fetus is fully formed at the end of the seventh month and other at the beginning of the ninth.
54. BT *Yevamot* 65b–66a.
55. BT *Yevamot* 66a, s.v., *achutei.* Both Judah and Chizkiah move to Palestine to become students of R. Judah HaNasi and both become revered for their sagacity.
56. BT *Yevamot* 63a–b.
57. Rosner (1971):93.
58. See Bleich's introduction in Rosner (1979):xix.
59. Bleich (1996b).
60. This story is found in JT *Terumot* 8:10. See also BT *Sanhedrin* 72b and Rashi there.
61. Bleich (1996b):101.
62. Bleich (2000):72.
63. This story is found, in part, in Tosafot, BT *Menachot* 37a, s.v., *o kum gelei.* Continuations of the story exist in Adolf Jellinek, ed., *Beit HaMidrash,* 4, Solomon, end; Gaster (1968): #113 in Hebrew (p. 75) and #392.11 in English (pp. 150–151); Ginzberg (1936) Volume 4:131–132.
64. Bleich (1996b):97.
65. *Shitah Mekubetzet,* BT *Menachot* 37a, #18.
66. Bleich (1996b):97.
67. Bleich (1996b):114n32. For a more thorough analysis of these stories regarding conjoined twins, see Crane (2012).

68. Rema at SA *Yoreh De'ah* 339:1. See Bleich (1996a):65, and 83n35.
69. Newman (2005):140.
70. See Crane (2009b) and Borowitz (2009).
71. Borowitz (1991):220.
72. Ibid.
73. Ibid., 219.
74. *Mekhilta deRabbi Ishmael, Yitro,* 5. See also *Mekhilta deRabbi Ishmael, Yitro* 6. Compare with *Sifrei Bamidbar, Shelach,* 115; *Sifrei Devarim, Ekev,* 38.
75. BT *Shabbat* 88a; *Midrash Tanhuma, Shoftim,* 9.
76. *Exodus Rabbah,* 5.9; *Midrash Tanhuma, Shemot* (Warsaw edition: #25; Buber edition: #22); *Yalkut Shimoni,* Job 921; *Deuteronomy Rabbah, Nitzavim* 5. See also *Mekhilta de Rabbi Ishmael, Yitro,* 5; *Sifrei Devarim,* 343.
77. BT *Shabbat* 88a.
78. Borowitz (1991):293.
79. Adler (1998):52.
80. Cover (1984):23.
81. Wimpfheimer (2011):18.
82. Newman (2005):141.
83. Ibid.
84. Cutter (1995):63.
85. He cried so loud when evacuating that even the grunts and brays of his farm animals could not cover his screams. See BT *Baba Metzia* 85a. The story of concern here is from BT *Ketubot* 104a. This translation follows the printed Vilna edition. Significant deviations found among other manuscripts have been noted. These include: Munich 95; St. Petersberg—RNL Evr. I 187; Vatican 113; Vatican 130; Soncino Print (1487).
86. Vatican 130 inserts here "the whole world came to."
87. St. Petersburg—RNL Evr. I 187 inserts that she heard ministering angels saying, "Rabbi's death is coming, Rabbi's death is coming."
88. Munich 95 omits her seeing Rabbi's frequency to the privy, and taking off his *tefillin* and putting them on. St. Petersburg—RNL Evr. I 187 also omits the frequency, removal, and donning of *tefillin;* but it adds that she saw him hurting greatly. Vatican 113 also omits the frequency, removal, and donning of *tefillin;* and its phrase is "when she saw that he was pained." Vatican 130 excludes reference to removal and donning *tefillin* and any indication that Rabbi was pained beyond needing to relieve himself frequently. She sees everything in Soncino Print (1487). Insofar as the detail about taking off and putting on prayer phylacteries is found only in the most recent (and printed) editions of the Talmud, we can conclude that this detail is a relatively late accretion to the story.
89. Vatican 130 insists that this was the whole world praying.

90. Vatican 130 says "the whole world."
91. The Erelim are a specific category of angels in the Judaic angelic pantheon. They are first mentioned in Isaiah 33:7 (see Rashi and Radak there) as crying angels. Their location among the other levels of angels is discussed in MT *Yesodei HaTorah,* 2.7; *Masekhet Azilut; Otzar HaMidrashim* (Eizenstein), *Gan Eden/Gehinom,* §86; *Zohar,* 2:43a/*Bo.* See also BT *Chagigah* 5b; *Bereshit Rabbah* 56:5; *Eikhah Rabbah* 1:2, §23; *Pesikta Zutra* (*Lekach Tov*), *Shemot* 33:22; *Zohar* 1:182a/*Vayeshev; Zohar* 2:195b–196a/*Vayekhel; Zohar,* 1:210/*Vayigash.* These angels are appointed over the grass, trees, fruits, and grain, and as soon as they have done the will of their Creator, they return to the place assigned to them and praise God; see *Otzar HaMidrashim* (Eizenstein), *Ketapuach Be'eitzah Hi'ar,* §262; *Sefer Orchot Tzadikim, Teshuvah,* §26.
92. Cutter (1995):66.
93. Ibid., 72.
94. Ibid., 74.
95. Ibid., 79, emphasis in original.
96. Cutter (1995):79.
97. Zoloth (1999):127ff.
98. Ibid., 194, emphasis in original.
99. Ibid.
100. Zoloth (1999):198.
101. Ibid., 218.
102. Ibid., 219.
103. Ibid. See summary of Zoloth's method in Newman (2005):142.
104. Newman (1998):197. See also Newman (2005):143. Jotkowitz (2012) also contends that narratives are part and parcel to modern Jewish medical ethics.
105. Fox (1975):4.
106. Wimpfheimer (2011):29.
107. See, for example, Bleich (1973); Sherwin (1974); Levine (1975); Raskas (1975); Bleich (1975); Dagi (1975); Bardfeld (1976); Leiman (1977).
108. Jotkowitz (2012).
109. Sharzer (2011):254.
110. Newman (2007):187.
111. Cutter (2006):59.

CHAPTER 3

1. Marquez (1982):50.
2. BT *Ketubot* 104a. See also BT *Nedarim* 39a–40b for discussions on visiting the sick and the efficacy of prayer for their healing. Many in the Jewish textual tradition prayed to die: Moses (Exodus 32:32;

Numbers 11:15), Jonah (4:3), Samson (Judges 16:30), Honi the Circle Maker (BT *Ta'anit* 23a). See the fourteenth-century scholar R. Nissim of Gerondi's commentary at BT *Nedarim* 40a for further insight about praying for death.

3. *Yalkut Shimoni*, 2:943.
4. BT *Gittin* 47b.
5. M *Avodah Zarah* 1.7 // BT *Avodah Zarah* 16a. Translations of rabbinic texts are mine unless otherwise noted. Biblical translations come from the *New JPS*.
6. A basilica was often used by Romans as a court for capital punishment; a scaffold (*geredom*) was a place of torture and execution; stadiums often hosted executions; a platform (*bimah*) was sometimes used to throw victims off to their deaths. On Amoraic restrictions of earlier Tannaitic prohibitions to assist gentiles in building such structures, see Hayes 1995.
7. Digital copies are available at http://www.lieberman-institute.com/.
8. Being sent to a brothel tent was also a conceivable punishment for (handsome) men; see *Midrash Tanhuma, Miketz*, §8; *Midrash Agadah* (Buber), *Bereshit*, 42, s.v., *vayirdu achi; Yalkut Shimoni, Miketz*, §148. Epstein (1935):181, argues that "professional harlotry" was prohibited even though concubinage (*peligesh*) was not. See MT *Na'arah B'Tulah* 2.17, and commentaries there. Elsewhere (BT *Gittin* 57b), it is unclear whether the children who were carried off for sexual impropriety were punished by Roman or Jewish authorities, and what they did to deserve this treatment; nonetheless, en route they committed suicide by jumping into the sea.
9. That these three people recited precisely these three verses is corroborated in *Sifre Devarim* §307, and partially in BT *Semachot* 8.11/47b (Soncino Edition). Though Abrams's (1995) claims that Beruriah, one of Chananya's daughters, recites (these) biblical texts here, this is difficult to uphold, as elsewhere it is the nameless daughter of Chananya who recites scripture over her brother's demise (BT *Semachot* 12.13; *Eikah Rabbah* 3.6) alongside her father's and mother's invocations—and not Beruriah. (While that now-dead son had associated with robbers (see text #67 in Gaster [1924]), Chananya's other son was a Torah scholar (Tosefta *Kellim* 4.9)). Moreover, the next *sugya* in the Talmud immediately following our central story begins with Beruriah asking Meir to release *her sister* from the brothel tent (BT *Avodah Zarah* 18a). Perhaps Abrams bases her claims on the late Talmudic comment by the *stam* (redactor), "There are [sages] who say [that Meir fled Babylon] because of the episode involving Beruriah," which Rashi explains was Beruriah's capitulation to an adulterous relationship and eventual suicide and Meir fled out of shame (BT *Avodah Zarah* 18b, *s.v., v'ika d'amrei*

mishum ma'aseh di'vrurya). Yet, the Talmud previously says the reason Meir fled is because he bribed the brothel guard to release his wife's sister—meaning Beruriah's nameless sister (BT *Avodah Zarah* 18a–b).

10. On acts of omission and being held accountable for others' acts of commission, see BT *Shabbat* 54b–55a; BT *Shevuot* 39a.
11. BT *Avodah Zarah* 18a, s.v., *'al kol darchei.*
12. Supposedly this brothel-consigned daughter walked seductively before Roman dignitaries. About her the dignitaries said, "How beautiful are the steps of this maiden," causing her to become even more conscientious of her gait. This immodesty violated the Psalmist's claim that the honor of the princess dwells within (Psalms 45:14). See BT *Avodah Zarah* 18a; Rashi at BT *Avodah Zarah* 18a, s.v., *dik'd'kah; Sefer Orchot Tzaddikim, Sha'ar HaGa'avah,* s.v., *ga'avah mit'ch'leket; Yalkut Shimoni, Tehillim,* §758.

    According to the Schotenstein translation of the Talmud, her being incarcerated in a brothel is fitting punishment. Since she excited men's libido (*yetzer harah*) with her voluptuous movements, now she will need to "resist her *yetzer harah* to participate in the activities of her housemates." The Talmud and this note express an insensitive androcentrism so profoundly afraid of women's beauty and male attraction to the female moving form, that the best way to preserve men's sexual integrity is to confine women and their sexuality to brothel tents. This note also assumes that this woman—if not all women—would be sexually stimulated by the thought if not sight and sound of nearby women being used for male sexual gratification.
13. Exodus 20:5; Deuteronomy 5:9.
14. Ezekiel 18:20; see also 18:4.
15. *Sh'niftar:* Vilna, Paris 1337, Munich 95, Pesaro 1515. *Sh'met:* JTS Rab 15, Jerusalem-Schocken 3654.
16. *G'dolei romi:* Vilna, Paris 1337, Pesaro 1515, JTS Rab 15. *B'nei romi* (the children of Rome): Munich 95. *G'dolei hador* (the elite of the generation): Jerusalem-Schocken 3654; #189 in Gaster (1924).
17. *L'kovru:* Vilna, Jerusalem-Schocken 3654, JTS Rab 15, Pesaro 1515.
18. *V'haspiduhu hesped gadol:* Vilna, Paris 1337, Munich 95, Pesaro 1515; #189 in Gaster (1924). The full phrase—*l'kovru v'haspiduhu hesped gadol*—is shared only between Vilna and Pesaro 1515.
19. *Yoshev v'osek:* Vilna, Paris 1337, Munich 95, Pesaro 1515, JTS Rab 15.
20. Besides its last word, Jerusalem-Schocken 3654 has an entirely different ordering, which is: Upon returning they found R. Chananya ben Teradyon with a Torah scroll resting on his lap, and sitting and convening substantial gatherings, and explicating (*v'doresh*). This version does not portray him "engaging in" Torah (*v'osek*) as do the others. Eisenstein's *Otzar HaMidrash, Asarah Harugei Malkut,* p. 244,

interpolates a comment and context: They said about Chananya ben Teradyon that he was a pleasant man before God and people, and never a disparaging word about fellows would arise from his lips. And when the Roman Caesar ruled that teaching Torah is prohibited, what did R. Chananya ben Teradyon do? He stood, convened gatherings, and sat in the Roman market where he taught and explicated the Torah. Caesar ruled that he be wrapped in a Torah scroll and burned.

21. *V'hatzito bahen et ha'or:* Vilna, Pesaro 1515., Paris 1337, Jerusalem-Schocken 3654, Munich 95; #189 in Gaster (1924). *V'hatzito bahen eish* (lit them afire): JTS Rab 15.
22. *K'dei shelo teitzei nishmato meheirah:* Vilna, JTS Rab 15, Jerusalem-Schocken 3654, Paris 1337, Pesaro 1515. #189 in Gaster (1924) does not include *meheirah. K'dei shelo yamut meheirah* (so that he will not die quickly): Munich 95.
23. *Aba:* Vilna, Pesaro 1515, Jerusalem-Schocken 3654, JTS Rab 15, Paris 1337; #189 in Gaster (1924). Missing in: Munich 95.
24. Rashi interprets this as follows: "This is the reward for [a life devoted to] Torah?!" A similar rhetorical question is uttered by ministering angels when R. Akiva is executed by iron combs raking through his flesh; see BT *Berachot* 61b. See also the Higer manuscript of BT *Semahot* 8.12 for a similar statement.
25. *B'iti* (my daughter) is found in Paris 1337, Jerusalem-Schocken 3654 preceding this phrase.
26. The first-person narrative is found in Vilna, Pesaro 1515, Paris 1337, Jerusalem-Schocken 3654. A third-person version, found in Munich 95, JTS Rab 15, goes as follows: *akshav sh'hu nesaref v'sefer torah 'imo* (now that the one who burns and the Torah scroll is with him).
27. *'Albonah.* This term, also understood as humiliation or insult, is associated with rejecting Torah study; see M *Avot* 6.2.
28. In Eisenstein's *Otzar HaMidrash, Asarah Harugei Malkut,* p. 244, Chananya's reply to his daughter is: It is good for me that you see me thus.
29. JTS Rab 15 reverses the order. Eisenstein's *Otzar HaMidrash, Asarah Harugei Malkut,* p. 244, adds: "And he began to cry. His students then inquire, for what reason do you cry? He said to them, If I alone were burning it would be a difficult thing for me, but now I burn and the Torah scroll is with me." The text then jumps to the conversation with the executioner.
30. *Af atah petach picha, v'tikanes [bechah] ha'eish:* Vilna, Pesaro 1515. *Petach picha v'yikanes shel'hevet beficha v'teitzei nishmatecha meheirah* (open your mouth and the flames will enter your mouth and your soul will depart quickly): Jerusalem-Schocken 3654. *Petach picha v'tikanes shel'hevet k'dei sh'teitzei nishmatecha meheirah* (open your

mouth and the flames will enter so that your soul will depart quickly): JTS Rab 15. *Af atah petach picha v'tikanes shel'hevet* (open your mouth and the flames will enter): Munich 95.

There is a play on the verb to rest (*lanuach*) in Paris 1337. The phrase at this point is: *Af petach picha v'tikanes bah ha'eish v'tamavet v'tanuach* (Also you can open your mouth and the fire will enter it and you will die and be at rest). This echoes the Torah scroll resting on Chananya's lap (*sefer torah moneach lo b'cheiko*), and the tufts of wool affixed to his chest cavity (*v'hanichum 'al libo*).

31. *Mutav sh'yitalnah mi sh'nitenah v'al yechavel hu b'atzmo:* Vilna, Pesaro 1515, JTS Rab 15, Munich 95; #189 in Gaster (1924). More will be said below about this phrase. On arguments about self-inflicted injury (*hachovel 'atzmo*), see BT *Baba Kama* 91b. As will be explained below, such injuries are not lethal.
32. *Kaltz'etoneiri:* Vilna, Pesaro 1515. *Kalisteneir[i]:* Munich 95, JTS Rab15, Paril 1337. According to Jastrow (1903), these spellings are corruptions of *Kostinar* (*quæstionarius,* torturer, executioner), as is found in Jerusalem-Schocken 3654.
33. *Marbeh b'shelhevet:* Vilna, Munich 95, Pesaro 1515, Paris 1337. *Marbeh b'shelhevet eish* (increase the flames of the fire): Jerusalem-Schocken 3654. *Marbeh lecha ba'etzim* (increase the timber for you): JTS Rab 15.
34. Though all versions are the same with this exchange, it is unclear who said what to whom. The conversation could be understood as follows: Chananya said, "Yes, swear to me [that you will do as you suggest]." He (the executioner) swore to him. The swearing is absent in text #189 in Gaster (1924).
35. JTS Rab 15 adds *lo* (for him): "he increased for him."
36. *Yetziah nishmato [bi]m'herah:* Vilna, Pesaro 1515, Munich 95. *V'yatztah nishmato* (his soul departed): JTS Rab 15, Paris 1337. This phrase is missing altogether in Jerusalem-Schocken 3654.
37. *Kafatz v'nafal l'toch ha'or:* Vilna. *Kafatz v'nafal l'toch ha'eish* (jumped and fell into the fire): Munich 95. *Kafatz l'toch ha'or v'nisaref* (he jumped into the fire and was burned): Paris 1337. *Kafatz l'toch ha'or v'yatztah nishmato* (he jumped into the fire and his soul departed): JTS Rab 15.
38. A parallel story of a Roman official seeking and negotiating for eternal life can be found at BT *Ta'anit* 29a. There, the Roman officer (*hegemon*) found Rabban Gamliel hiding from the decree against him. The officer said to him, "If I save you from your death, will you bring me to the World to Come?" Gamliel replied, "Yes." He said, "Swear to me." And he swore to him. The officer then ascended to the attic, threw himself down, and died. A *Bat Kol* then came forth and declared that the officer has been assigned to life in the World to Come (*mezuman l'chayei ha'olam habah*).

39. The same closing lament is found elsewhere (BT *Avodah Zarah* 17a) in regard to Elazar ben Durdia, who was tempted by carnal relations with every earthly harlot, and yet when he ultimately resisted the last one and repented, he died. And elsewhere (BT *Avodah Zarah* 10b) it is also related to Ketiah bar Shallum, a righteous Roman minister, who defended the Jews against a genocidal Caesar's plan to eliminate the Jews because they were, in his eyes, like a wart needing excision. Ketiah was put into a round furnace or chamber for besting Caesar in argument. He took off his own foreskin so that he would merit the World to Come without passing through Gehinnom. Eisenstein's *Otzar HaMidrash, Asarah Harugei Malkut*, p. 244, inserts "like this executioner" for the first part, and to the latter "and there is a person who worships Adonai all the days of his life and suffers his reward in one moment, like Yohanan the High Priest, who served 80 years as the High Priest, and whose end was vindicated (*ul'b'sof na'aseh tzaduki*)."
40. According to Joseph Karo in his Shulchan Aruch, *Orech Chayim*, 580.2, this occurred on the 27th of Sivan (May–June). R. Eliezer Yehuda Waldenberg offers a lengthy justification for this calculation. See *Responsa Tzitz Eliezer* 15.10.
41. M *Sanhedrin* 7.2. See also BT *Sanhedrin* 52a–ff.
42. Paris 1337. The Jerusalem-Schocken 3654 reads, "It is better that the one who takes it is the one who gave it in me. Do you hold that I should injure myself?" (*mutav sh* <..> *lana mi sh'nitenah bi memah sh'echavel ani b'atzmi*).
43. This is picked up by the thirteenth-century Spanish exegete, R. Yom Tov ben Ishbili (HaRitba). He heard that in France it was permitted for individuals to injure themselves lest they fear being compelled to transgress their religion. From this story, and in conjunction with other textual precedents, he rules that self-injury is a positive obligation under such dire circumstances. See *Chidushei HaRitba, Avodah Zarah* 18a.
44. See note 38 above.
45. This story follows the Finkelstein edition, published in New York (1968). Variants are noted as needed.
46. Even though the text says that he is to be burnt with his scroll (*'im sefro/sefarcha*)—meaning his Torah scroll—Basser (1990) argues that the original story of Chananya's death incorporated no Torah scroll. All such references are later accretions to the *Sifrei Devarim* version. This argument is supported by the fact that his wife and daughter are informed that his punishment is to be burnt, but no scroll is mentioned. And text #67 in Gaster (1924) corroborates this pattern. On the other hand, the version found in *Yalkut Shimoni, Parashat Ha'azinu*, 942, incorporates the scroll in every instance.

47. Chananya offers only the first quarter of the verse. The rest of the verse reads, "For all [God's] ways are just; a faithful God, never false, true and upright is [God]" (Deuteronomy 32:4).
48. As noted above, only in the *Yalkut Shimoni, Parashat Ha'azinu,* 942, a mention is made of him burning with his scroll. The Hebrew does not specify that it was a Torah scroll, however, yet this could be surmised given Chananya's profession.
49. *Malachah.* According to Reuven Hammer (1986:494n14), the author of this *piska* is using "polite language," in contrast to the vulgarity of the Talmud, which states that the daughter is condemned to dwell in a brothel tent (*l'eisheiv b'kuvah shel zonot*); see BT *Avodah Zarah* 18a. This more explicit phrase is also found in *Yalkut Shimoni, Parashat Ha'azinu,* 942. Yet text #67 in Gaster (1924) says she is assigned to work on Shabbat (*la'asot mal'achah b'shabat*).
50. *Gedolim tzaddikim elu. Pitron Torah, Parashat Ha'azinu,* page 310 (Orbach edition, published in Jerusalem, 1975) omits *tzaddikim,* as well as the following phrase about their uniqueness in all scripture. In *Yalkut Shimoni, Parashat Ha'azinu,* 942, this phrase is *gedolim ma'asim elu* (these are great deeds).
51. *Eparchi.* This would be a prefect or governor of a province or town. See also BT *Shavuot* 6b; *Bereshit Rabbah* 11.4.
52. *Al tazuach da'atcha.* In *Pitron Torah, Parashat Ha'azinu,* page 310, this phrase is *al tashbiach 'atzmecha* (do not aggrandize yourself).
53. The philosopher in *Pitron Torah, Parashat Ha'azinu,* page 310, offers a dramatically different rationale for not being haughty: Because [Torah] was given in fire and it returns to fire; and God, the Holy One, will exact lethal punishment from you on account of these righteous ones.
54. On this phrase, see also *Bereshit Rabbah, Vayeshev* 88.5.
55. This phraseology echoes that of R. Jose at BT *Avodah Zarah* 18a.
56. Basser (1990):71, asserts that the *Sifre* version is older than all other versions and serves as their source. He surmises this from the fact that this text does not describe the death scene or offer precise reasons for the suffering of Chananya and his family (p. 73). In his view, later texts fill in these details. It is unclear to me, however, why later texts, like the Talmud, would excise this conversation between the philosopher and governor and put another—concocted—conversation in its stead. If Basser is correct and this were the case, then the argument that I will make below stands on more solid ground: this text is shaky, at best, upon which to establish norms.
57. See discussion of dating of *Semachot* in Zlotnick (1966):1–9; he surmises that its provenance is closer to the early end of the range. On the dating of *Kallah,* see Brodsky (2006):34–84. According to his close analysis, this text was redacted somewhere between the fifth

generation of tannaim and the fourth generation of ammoraim, that is, between the second-century CE and fourth-century CE. Basser (1990):80, states, "These minor tractates are to be dated to the dawn of the Amoraic period and may essentially be considered as Tannaitic texts."

58. The translation here follows the Soncino edition. This crime is also mentioned in BT *Avodah Zarah* 17b and 18a.
59. *Yoshev v'tameh.* Only a few other significant characters sit in confusion: Abraham (*Pesikta Rabbati,* §15; *Yalkut Shimoni, Lech Lecha,* §77); Moses (*Yalkut Shimoni, Beshalach* §233); Korach (*Torat HaMinchah, Shelach-Korach,* 9, page 529).
60. *Nitchayavti mitah:* According to the R. David Kimchi (Radak), Moses also thought he was obliged to die. See Radak's commentary on Psalms 3:1. See also *Yalkut Shimoni, Ovadiah,* §549.
61. *Kostinar: quæstionarius,* torturer, executioner.
62. *K'corchecha u'lsorefcha b'toratecha.* Basser (1990) astutely observes that the text does not say *sefer Torah* but only *Torah*—reflecting a rabbinic tradition of seeing Chananya as a living Torah.
63. *U'lyisrael imecha.* The Higer edition, however, reads: *b'toratecha u'lsorefcha imah v'amad.*
64. A fire not burning its human captive is rare. When Nimrod threw Abraham into a fiery furnace, he was protected by none other than God, even though Gabriel asked for the privilege to do so (BT *Pesachim* 118a; see also versions of this story in Qu'ran 21:51–70; *Bereshit Rabbah* 38.13). Hananiah, Mishael, and Azariah were also thrown into a fiery furnace by Nebuchadnezzer, and though Yurkami, the angelic master of hail, requests the privilege of saving them, it is Gabriel, the angelic master of fire, who gets that honor (BT *Pesachim* 118a; *Midrash Tehillim* 117:3; *Yalkut Shimoni, Tehillim,* §873; *Sefer Ha'Ikkarim* IV:6). The significance here is that Chananya does not require a direct divine intervention to be saved from the fire; he himself suffices and serves as a retardant.
65. *Reikah.* This epithet is also said by R. Eleazar ben Shimon about an exceedingly ugly man (*mechu'ar me'od*) (BT *Ta'anit* 20b).
66. Such animals are considered so lethal that when involved in a crime against humans, they are to be brought to a full court of 23 rabbis and are to be executed (see M *Baba Kama* 1.4; M *Sanhedrin* 1.4). Added to this list are panthers and leopards. In *Avot d'Rabbi Natan,* 1.40, all these animals are understood to both see and be seen. Whereas these other texts speak of animals per se, it seems Chananya speaks of these animals as if they were analogues to other tyrannical human regimes like Rome.
67. A similar—nearly verbatim—claim that the designated victim will die regardless of what the government does, can be found at *Eikah Rabbah* 1.50.

68. The Higer edition reads: *kafatz v'nafal l'toch ha'or v'ashmi'a kolo m'toch ha'esh*—he jumped and fell into the light and made his voice heard from within the fire.
69. This is verbatim with what is said in the *Avodah Zarah* version.
70. See *Nachalat Ya'akov*'s commentary on this point, *s.v., kum lach,* as found in the Soncino edition.
71. See, for example, Rashi at Ruth 1:16.
72. BT *Semachot* 8.11/47b (Soncino edition). In the Higer edition, it is 8.12.
73. *U'k'sh'nicanes:* and when he was gathered, according to the Soncino. *U'k'sh'nitpas:* and when he was arrested, according to the Higer version.
74. *L'minut.* This is also recorded in text #289 in Gaster (1924). Heresy was R. Eleazer ben Hyrcanus's crime, too; see BT *Avodah Zarah* 16b.
75. *S'reifah:* Soncino, Higer. Michael Rodkinson (1903) seems to have had a manuscript that said he was to be killed at the stake; see his BT *Semachot* 8.7.
76. *l'yeshev b'kufah:* Soncino. Higer, however, is more explicit: *l'yeshev b'kuvah shel zonot*—to dwell in a tent of prostitution.
77. *Harigah:* Soncino, Higer. Rodkinson says, "To the sword"—meaning she was to be decapitated, which may have been what *harigah* meant in Roman terms.
78. The Higer edition continues the quote to include, "and loving toward all [God] made." According to Higer, Chananya also says, "The Rock! [God's] deeds are perfect, all [God's] ways are just. A faithful God, never false, true and upright [is God]" (Deuteronomy 32:4)—precisely what Chananya and his wife recite in the BT *Avodah Zarah* 18a version.
79. While Soncino quotes only these words, the Higer recites the full verse, "Wondrous in purpose and mighty in deed, whose eyes observe all the ways of men; so as to repay every man according to his ways, and with the proper fruit of his deeds." This verse is what the daughter recites in BT *Avodah Zarah* 18a.
80. Higer adds: *v'sefer torah 'imo*—and a Torah scroll with him. The second half of text #67 in Gaster (1924) offers a slightly different version: the Torah scroll is not explicitly mentioned (yet).
81. Higer adds: *v'amrah zoh torah v'zoh s'charah*—and she said, "This is Torah and this is its reward?" Rashi offers a similar sentiment at BT *Avodah Zarah* 18a, s.*v., eracha b'chah.*
82. *Mutav sh'tochalni eish sh'nufchah v'lo eish sh'lo nufchah.* According to Rodkinson, "is it not better I should be consumed by a fire which was kindled in this world than by a fire which is not kindled (Gehenna)?"
83. The whole verse reads, "Utter darkness waits for his treasured ones; a fire fanned by no man will consume him; who survives in his tent will

be crushed." The whole chapter is a meditation of what will befall the wicked.

84. In text #67 in Gaster (1924), the daughter says, "I cry for the Torah that is burning with you."
85. In text #67 in Gaster (1924), a proof-text is interpolated here: *mimino eish-dat lamo* (Deuteronomy 33:2). It is unclear what this means. According to some it means a law of fire flashes from God's right; to others it means lightning flashes at them from God's right; to some it means from the south [God came] to them at the slope; and to some, from the southland You proceeded to them. See discussion in Tigay (1996):320.
86. *Harei otiot porchot v'ein ha'eish ochelet ela 'al hanayer:* Soncino. The Higer reads: *harei hen haketuvin porchin ba'avir, v'ein ha'eish ochelet ela ha'ur bilvad*—behold, thus what is written soars in the air, but the fire consumes the parchment only. In Rodkinson, "The parchment only is burned, but the letters fly away." In text #67 in Gaster (1924): "The skin is burning, but the letters soar and they are not burning."
87. Rodkinson has, "Thou must also know that the great servants of the king are mostly beaten through the lesser," and then the verse. The rest of the verse reads, "And the day that dawned [brought on] your punishment," which Basser (1990):69, takes to mean that "the death of the prophet [Hosea comes] with the burning of the scroll."

    The Higer edition places this comment about great servants being cut down by lesser ones not here (8.12) but at 8.10, which has a parable about guests who linger too long at a king's banquet. Nonetheless, Basser (1990):70, believes that all stories associated with Chananya's martyrdom "were always associated with Hosea 6:5." This argument is weak at best insofar as this is the only version of his demise that makes reference to this particular prophet or statement. It could very well be that the inclusion of this verse here in the Soncino edition is a mistake, and that it rightfully belongs where Higer places it—next to a story about a king and his guests.

    Text #67 in Gaster (1924) offers a conclusion more akin to that found in BT *Avodah Zarah* 18a. Chananya says to his daughter: who in the future will address the affront of the ruling of the Torah will [also] address the affront of my blood from their blood (*mi sh'atid lifro'a dinah shelatorah yifro'a damay midam*). A moderately similar phrase occurs in *Eikah Rabbah* 1.50.
88. MT *Teshuvah*, 3:6–8.
89. BT *Rosh Hashanah* 17a. See also *Shemot Rabbah* 19.4.
90. See Job 20:1, 26, 29.
91. Scrolls written by heretics are to be destroyed; BT *Gittin* 45a. This applies to phylacteries and *mezzuzot* as well. If such texts were in a heretic's possession, they are to be hidden away. See also SA *Orach*

Chayim 39.1, *Yoreh De'ah* 281.1. Elsewhere the Talmud says that wise teachers are to be burnt with their scrolls (BT *Gittin* 58b).

92. See also Deuteronomy 33:2; Jeremiah 23:29; BT *Ta'anit* 7a; *Devarim Rabbah* 3:12; *Midrash Tanhuma, Bereshit* 1; Zohar, *Shemot*, 2/84a.
93. The Tosafists of medieval France opine that Chananya's crime was not capitulating to worship idols. See Tosafot at BT *Avodah Zarah* 3a.
94. Finkelstein (1938):47, claims that after hastening Chananya's death, the executioner "was himself condemned to death on the next day for this act of pity." This claim curries no support in any of the versions mentioned by Finkelstein (see note 17 there) or studied here, however.
95. Unless one considers the protests of the philosopher (B) and the daughter (D) as interventions. This fails, however, insofar as the philosopher seemingly speaks after Chananya had already been burned or currently is being burned elsewhere and thus no intervention is possible, and the daughter expresses anguish at witnessing her father's burning flesh but offers no concrete alternative.
96. See, for example, Basser (1984):100n23.
97. Wimpfheimer (2011):149.
98. Increasing attention has been given recently to the connection between ambivalence and ambiguity in rabbinic texts. See, for example, Belser (2011); Crane (2011b); Wimpfheimer (2011).

CHAPTER 4

1. Kierkegaard (1959):151.
2. Moshe Idel notes a peculiar alteration of the Talmudic detail about Chananya's condition of being surrounded by twigs and burned. The *Eleh Ezkarah* prayer says *u'vachavilei z'morot sarfu galmo*, which Idel translates as "they burned his body using bunches of branches." The word *galmo* is a concept (i.e., *golem*) that takes on mystical meanings in other medieval literature. See Idel (1990):285.

   For a modern reinterpretation of this prayer, see the one by Rabbis Melissa Weintraub and Sheila Peltz Weinberg at http://www.rhr-na.org/issuescampaigns/torture/resources/138-eleh-ezkerah.html, accessed February 6, 2012. Note this version's conscientious segmentation and alteration of the story of Chananya's demise.
3. For a discussion of the diversity among these lists, see Finkelstein (1938).
4. After reviewing classic sources, Rosner (1998):76, opines that Jewish law defines suicide as "one who had previously announced his intentions and then killed himself immediately thereafter by the method he announced."
5. See Droge and Tabor (1991).

6. In a different version of his final moments, Saul pleaded to be killed by an Amalekite—who acceded to his request (I Samuel 1:1–10). For a discussion of Saul's actions, consider *Bereshit Rabbah* 34.13. See also *Pirkei d'Rabbi Eliezer* 33; *Vayikra Rabbah* 26.7.
7. According to *Targum Yonatan* at Judges 16:28, Samson prays for strength to complete his lethal task. See further analysis in Crane (2008).
8. BT *Gittin* 57b. That women, at least, would rather die than be violated, see BT *Ketubot* 3b.
9. BT *Ta'anit* 29a.
10. After waking from a 70-year nap, Honi HaMegagel prayed to die after realizing his friends and colleagues were dead. From his death Rava taught: either companionship or death. See BT *Ta'anit* 23a.
11. See, for example, MT *Rotze'ach U'Shmirat Nefesh* 2.2.
12. See, for example, BT *Semachot* 2.1; SA *Yoreh De'ah* 354; Novak (1974):90–93.
13. *Mutav sh'yitalno mi sh'notnah, v'al yishlot hu b'atzmo.* See his *Yoreh De'ah* 326 and *Even Ha'Ezer* I:69. See discussion in Goldstein (1989):86ff.
14. Novak (2007):124.
15. Leiman (1977):45.
16. Novak (1974):84. Though Novak consistently refers to this Talmudic story of Chananya's death in his scholarship, he shifts how he applies it. Whereas in 1974 he invokes the story to illustrate his notion of martyrdom, in 1976 he uses it to critique Immanuel Jakobovits's mid-century argument that this story supports the concept of removing impediments to death, and in 2007 he employs it when deliberating the case of a physician ordering a patient to commit suicide.
17. See discussion in Van Henten (2004):164–165.
18. Finkelstein (1938):32.
19. *Bereshit Rabbah* 64.10.
20. BT *Ta'anit* 18b. For a parallel rebuttal, see *Eikhah Rabbah* (Buber), 84–86.
21. Finkelstein (1938):41.
22. BT *Sanhedrin* 74a. Rabbenu Nissim (ibid.) of fourteenth century Spain disagrees. In his view, it would be better for Jews to be killed "so that one letter of the Torah not be changed."
23. Kravitz (2006):89.
24. Rosner (1998):74.
25. Rosner (2006):348.
26. Goldstein (1989):56.
27. See BT *Sanhedrin* 45a, 52b; BT *Pesachim* 75a; BT *Sotah* 8b. At BT *Ketubot* 37b, this stipulation extends even to the killing of a heifer that atones for a murderer.

28. See critique of modern invocations of this concept in Berkowitz (2006):31ff.
29. Freedman (1999):286; Knobel (2007):177.
30. See Crane (2011a, 2011c).
31. BT *Avodah Zarah* 17b–18a.
32. Blidstein (1984):57.
33. Ibid.
34. According to the Talmud: his wife merited death for not restraining him from speaking so in public; his daughter deserved the brothel because she once responded to a compliment about her gait by walking even more carefully.
35. For a fuller recapitulation of the *sugya*, see Chapter 3.
36. Blidstein (1984):57.
37. Ibid., 61.
38. See, for example, the introductory comments to this prayer in Silverman (1986):381–382.
39. Basser (1990):71.
40. Berkowitz (2006):207.
41. Ibid. It is unclear why Berkowitz claims his family also receives placement in the World to Come, since no textual source attests to this.
42. Burning was one of the four acceptable methods of execution; the others were stoning, beheading, and strangulation. See M *Sanhedrin* 7.1.
43. M *Sanhedrin* 7.2. R. Judah disagrees with the rope method; he urges the use of tongs to pry open the individual's mouth.
44. Berkowitz (2006):167.
45. BT *Berachot* 61b. The following translation follows the printed edition. Departures found among manuscript variants are footnoted.
46. This question is missing in Oxford Bodleian 2673 (hereafter Oxford 2673).
47. This detail exists in Soncino Print (1484), Munich 95, Vilna, Oxford 2673. It is missing from the Oxford Bodleian Additional folio 23 (hereafter Oxford 23), Paris 671.
48. Oxford 23 and Paris 671 add that he was preparing his mind to receive the yoke of Heaven with love. Oxford 2673 also ends with "with love."
49. *Mitsta'er*—to be troubled. Some manuscripts have him say, "All the days of my life I used to interpret (*doresh*) this verse . . . ": Paris 671, Oxford 23.
50. *Notel.* This is the same root as that expressed in Chananya's moral, as found in *Avodah Zarah* (see Chapter 3).
51. Some manuscripts have a version that states something akin to: he did not break to complete the statement until his soul departed with *echad*. These include Oxford 2673, Munich 95, Oxford 23, Paris 671.

52. This is missing in Munich 95, Oxford 23, Paris 671. It is found only in Soncino Print (1484), and Vilna.
53. This completes the phrase quoted by the angels of Psalms 17:14. All manuscripts have God speaking these last two words, except in Oxford 2673 God does not appear at all—the angels complete the quote.
54. In Oxford 2673 the *Bat Kol* combines both statements: Fortunate are you, R. Akiva, for your soul went out with *echad,* [and] that you are assigned to life in the World to Come.
55. A completely different version of Akiva's last moments is available in JT *Sotah* 5.5/25a–b.
56. If one takes seriously the existence of *Avodah Zarah* manuscripts Paris 1337 and Jerusalem-Shocken 3654.
57. On the other hand, *someone* replied to Chananya's inquiries about his family members' punishments. Though it is reasonable to surmise that his interlocutors were Roman authorities, they could very well have been Jewish because he asked of them, "What have *they* decreed" and not "what have *you* decreed," and the response he got was not in the first person plural but in a passive voice.
58. Kierkegaard (1959):151.
59. There is, however, an intriguing mystical variant of the Chananya story in which the ruling magistrate is identified as Lupinus Caesar. In this version, Surya, the Prince of the Presence, was commanded by God to substitute Lupinus and Chananya while they slept. So when it came time to execute Chananya, the person who looked like Chananya was actually a Roman leader—and he was swiftly decapitated. Chananya, appearing like Caesar, ruled ruthlessly, killing pagan priests by the thousands each month, as if to purge Rome of its wayward theo-political institutions and inclinations. Caesar somehow re-emerged in the guise of Chananya, was captured, and burnt alive. He was then reconstituted in the heavens and brought before a heavenly court, which ultimately cast him into a fierce fire that caused him unimaginable anguish. See *Hekhalot Rabbati,* VI:117–120; Smith (2009).
60. M *Sanhedrin* 10.1–4 discusses who among all humans and even within the community of Israel are eligible for the World to Come. See also BT *Sanhedrin* 105a; T *Sanhedrin* 13.1; JT *Pe'ah* 1.1; MT *Teshuvah* 3.4. See the thorough discussion of these and many other sources in Novak (1983).
61. The following translation follows the printed edition of BT *Ta'anit* 29a. Manuscript variants include Munich 140, Munich 95, Oxford 23, Vatican 134, Cambridge T-S AS 79.67, Goettingen 3, British Library 5508 (400) (hereafter BL 5508).
62. This was most likely Q. Tineius Rufus, an imperial legate, who briefly served in Judea (~129–132 CE). See discussion in Katzoff (1993):142.

63. Other manuscripts read burned (*saraf*): Goettingen 3; BL 5508; Munich 95; Vatican 134. One version says ploughed (*charash*): Oxford 23.
64. *Heichal:* Vilna. *Ha'ulam* (hall): Goettingen 3; Munich 140; Munich 95; Oxford 23; Vatican 134. Munich 95 and Vatican 134 add that in addition to the hall being destroyed, the city and its seed were to be silenced.
65. *Adon:* Vilna. *Hegemon:* Goettingen 3; BL 5508; Munich 140; Munich 95; Oxford 23; Vatican 134.
66. Vatican 134 stresses that he made this announcement at the opening (front door) of the House of Study. Goettingen 3 insists he merely came to the House of Study.
67. It is considered that the Roman confused *nasi,* the Hebrew term for the president of the Sanhedrin, with the Latin *nasi,* which means "of the nose." See Katzoff (1993):141n3.
68. Vatican 134 inserts *rabu* here, suggesting that this was either a rabbinic tradition or a pervasive one among Romans. Following Katzoff (1993), it would be more logical to refer to Roman magistrate authority practices.
69. Goettingen 3 adds, "Happy are you, officer, for you are assigned . . . "
70. *Adon:* Vilna. *Hegemon:* Goettingen 3; BL 5508; Munich 140; Munich 95; Oxford 23; Vatican 134; Cambridge T-S AS 79.67.
71. All have *mezuman,* except Munich 95, which reads, "it's been decided" (*metukan*).
72. See Katzoff (1993) for more details about the temporal nature of Roman magisterial edicts.
73. BT *Baba Metzia* 59b.
74. See, among others, Luban (2004); Greenwood (1997); Stone (1993).
75. See, for example, Luban (2004):204. One obvious logic trap of this story is that the majority here is incorrect; it is Eliezer's singular position that accords with divine will. This therefore counters the biblical stipulation that R. Jeremiah perverts by not quoting the verse in its fullest: "*do not follow the majority to do wrong*" (Exodus 23:2).
76. *Kol hamoser 'atzmo l'mavet 'al divrei torah ein omrim halakhah mish'mo.* BT *Baba Kama* 61a.
77. Gross (2005):15–16n39.
78. van Henten (2004):180.
79. Droge and Tabor (1992):102. This suggests that the textual variation Droge and Tabor were working with had a *mutav* statement that read "one should not" and not "I will not." Strangely and inconsistently, they translate that very statement as, "Let him who gave me [my soul] take it away, but no one should injure himself" (101). Yet in their question quoted above, they interpret Chananya's teaching against self-*injury* as a rule against self-*destruction.*

80. Blidstein (1983):57; Gross (2005):15n38. However, the medieval scholar Judah ben Samuel (twelfth century) of Regensburg, in his *Sefer Hasidim*, §776, insists that Chananya wanted to be killed.
81. Goldstein (1989):28.
82. See discussion in Dorff and Crane (2012) and Newman (1998).
83. Kierkegaard (1959):222.
84. See Paris 1337 and Jerusaelm-Schocken 3654 manuscripts of BT *Avodah Zarah* 18a.

Chapter 5

1. Ricoeur (1992):161.
2. Gazzaniga (1998):1–2.
3. MacIntyre (1984):215.
4. Ibid.
5. See Ricoeur (1992):161, and see his critique of MacIntyre there.
6. See Chapter 1 for a historical sketch of modern Jewish bioethics, and Chapter 2 for the emerging role of narratives therein.
7. Jakobovits (1959):122. For an analysis of Jakobovits's methodology and its apparent responsiveness to the then pervasive Catholic bioethical deliberative processes, see Gray (2009).
8. Bettan (1950):124.
9. See, for example, the entry under Martyrdom in the 1901–1906 *Jewish Encyclopedia* (www.jewishencyclopedia.com), and after that, see Julius Preuss's 1911 *Biblisch-Talmudische Medizin*, wherein this source is listed under Suicide (Preuss [1993]:516). In *The Universal Jewish Encyclopedia* of 1941–1948, the entry on Chananya mentions this incident and insists that his observation about soaring letters "has often been quoted as a restatement of the Jew's faith in the ultimate triumph of the spirit" (Volume 5:201).
10. This is also noted but not expanded upon by Dorff (1998):408.
11. Inquiring about seeing a text, however, is not new. The prophet Zechariah is asked by his angelic guide about what he sees, and Zechariah replies, "a flying scroll, twenty cubits long and ten cubits wide" (Zechariah 5:2). See discussion in BT *Gittin* 60a; *Vayikra Rabbah* 6.3. Compare with other prophetic instances of being asked "what do you see?"—Zechariah 4:2; Amos 7:8, 8:2; Jeremiah 1:11.
12. Prouser (1997):14. Compare with Sherwin (1974):20.
13. Resnicoff (2003):88; Goldstein (1989):28; Sherwin (1998):83.
14. Shulman (1998):194.
15. Jakobovits (1959):123.
16. See, for example, Knobel (1995):43.
17. Breitowitz (nd). In his *Ye Shall Surely Heal: Medical Ethics from a Halachic Perspective* (1995), Yaakov Weiner translates the phrase as, "It is better that the One Who gave the soul should take it"—found in Resnicoff (1998):n93.

18. Green (1999):37.
19. Bleich (1996):83n35.
20. Rosner (1998):67.
21. Jacob (1985).
22. Biblical instances of praying for one's own death: Moses at Exodus 32:32 and Numbers 11:14–15; Samson at Judges 16:30; Elijah at I Kings 19:4; Jonah at Jonah 4:8. Job's wife may also encourage praying for death (Job 2:9–10). See also Honi the Circle Maker at BT *Ta'anit* 23a. In regard to praying for someone else's death, or at least associating prayer with another's death, see: R. Yehudah Hanasi's handmaid at BT *Ketubot* 104a; R. Yohanan's colleagues at BT *Baba Metzia* 84a; R. Ada bar Ahava fasting for his son's death at YT *Shabbat* 19:2/87b; and R. Yosi ben Chalafta granting permission to a morose old woman to cease praying, at *Yalkut Shimoni*, Proverbs §943, and Deuteronomy §871. Nissim of Gerondi rules that sometimes it is necessary to pray for someone's death (Ran at BT *Nedarim* 40a); this is upheld in *Aruch HaShulchan, Yoreh Deah* 335.3; *Iggeret Moshe, Choshen Mishpat* II:74D; *Iggeret Moshe, Choshen Mishpat* II:73a; *Nishmat Avraham*, 335.5. See Crane's "Medical Futility and Liturgical Efficacy: when clinicians pray for patients to die."
23. Reisner (2000):243. See also Prouser (1997):14; Resnicoff (2003):88. Weiner (1995):24 reads, "I myself should not wound myself."
24. Novak (1974):84. See also Novak (2007):123; Finkelstein (1938):46.
25. Glick (1999):49–50.
26. See CCAR (1950); Rosner (1970):321; Rosner (1986):54; Rosner (1998):67; Shulman (1998):194; Rosner (2006):352. Though without speaking of accelerating one's death, Novak (1976):105 offers, "Let Him who gave life take it, and let not a man kill himself." See also Hochhoizer (1992):112.
27. Goldstein (1989):81. This differs from his other translations at pages 28, 37, 89. Brody (2003):233 reads, "I should not kill myself."
28. Preuss (1971):607; Preuss (1993):516.
29. See, for example, the eighth chapter of *Baba Kama* (BT *Baba Kama* 83b–93a); MT *Chovel u-Mazzik;* SA *Choshen Mishpat,* 420–424 (under the category *chovel bachavero*).
30. See, for example, Jakobovits (1959):123; Preuss (1993):516; Barilan (2003):143; Brody (2003):233. The manuscripts of BT *Avodah Zarah* 18a that do have such statements are Paris 1337 and Jerusalem-Schocken 3654. See Chapter 3.
31. Reisner (2000):243. See also Novak (1976):105ff; Rosner (1998):61; Prouser (1997):14–15, n64; Goldstein (1989):38; Brody (2003):233.
32. Weiner (1995):24. See also Resnicoff (2003):88.

33. Brody (2003):238.
34. Novak (2007):125.
35. Shulman (1998):194–195.
36. Prouser (1997):14.
37. *Responsa Tzitz Eliezer*, 5, *Ramat Rachel*, §28. See also Jotkowitz (2012).
38. Sherwin (1974):20.
39. Jakobovits (1959):122–123.
40. Weiner (1995):24–25.
41. Dorff (1998):181; Dorff (1999):266.
42. Telushkin (2009):382, 383.
43. Of those who stop their readings of the *sugya* with Chananya's teaching to his students, see: Resnicoff (2003):88; Meier (1986):54; Prouser (1997):14. Though Hochhoizer (1992):112 recites the vast majority of the *sugya*, she does stop her analysis of it at the *mutav* statement. Barilan (2003) recites the whole *sugya* but does not delve into all its aspects, especially beyond the *mutav*.
44. Rosner (1998):67.
45. Ibid. The story referred to is BT *Ta'anit* 29a. See Chapter 4 for an analysis of that story.
46. BT *Baba Kamma* 61a. This is cited at Rosner (1998):69.
47. Rosner (1998):76.
48. These strategies may take their cue from the 1906 *Jewish Encyclopedia* in which the entry on Hananiah (Hanina) b. Teradion concludes its summary of this story with, "Thereupon the executioner removed the wool and fanned the flame, thus accelerating the end, and then himself plunged into the flames."
49. Sherwin (1974):20.
50. Green (1999):37.
51. Ibid., 38. See Sherwin (1974):21.
52. Bleich (1996):83n35.
53. Ibid.
54. Steinberg (2006):338.
55. Ibid.
56. MT *Mamrim* 2:4; Elon (1994):Volume II:519–520, 533–536; Yuter (2001).
57. This is true for biblical rules, to be sure. Rabbinic legislation may, in certain circumstances, be subordinated to enduring exemptions. See MT *Mamrim* 2:9. Yet the paradox persists: a rabbinic authority cannot permanently declare another rabbinic authority's ruling to be exceptional, for that very declaration is exceptional.
58. Novak (2007):124.
59. Ibid.
60. See Abraham (2006):370; Sherwin (1998):86; Herring (1984):80, 83; Rosner (2006):352.

61. Dorff (1999):266; Barilan (2003):158.
62. Shulman (1998):194.
63. Ibid., 194–195.
64. Rosner (2006):352.
65. Weiner (1995):25.
66. Ibid., 30.
67. Ibid., 31.
68. BT *Yoma* 85a; SA *Orech Chayim* 329:4.
69. Weiner (1995):31.
70. BT *Kiddushin* 42b; *Baba Kama* 79a, 87a; *Baba Metzia* 10b. *See also* MT *Sheluchin v'Shutafin* 1:1; Tur, *Choshen Mishpat* 182:4. See discussion in Crane (2010):340–341.
71. Telushkin (2009):383.
72. Ibid.
73. Kravitz (2006):81. Barilan (2003):158, "Hanina's refined discrimination between opening his mouth and paying the executioner to remove the wool is contrasted by the very existential choice of life and death." Though it is patently unclear where Barilan gets the idea that Chananya *paid* the executioner, it could be surmised that he uses this term to refer to the quid pro quo nature of the exchange between them. This barter, I suppose, could be understood as a kind of payment, but construing it as such for a modern audience hides the fact that it was not a pecuniary transaction but a theological one.
74. See Novak (2005):33. Novak distinguishes contract from covenant.
75. T *Sanhedrin* 13:2.
76. Knobel (1995):43. See also Knobel (2007):177.
77. Green (1999):37.
78. Ibid., 38. See also Brody (2003).
79. Green (1999):38.
80. Kravitz (2006):93.
81. Prouser (1997):14.
82. Herring (1984):74.
83. Weiner (1995):24, 39.
84. Resnicoff (2003):88–89. Some might point to BT *Berachot* 5a as rabbinic proof of this position. Resnicoff does not, however; he instead points to fellow contemporaries who proclaim by fiat that suffering is no warrant for escape.
85. Kravitz (2006):81.
86. Ibid.
87. Ibid.
88. Ibid., 93.
89. Dorff (2000):326n3; see also p. 311; and Dorff (1998):199. It is surprising that he nonetheless invokes this text in his own arguments about end-of-life decisions and care.

90. Newman (1998):256n15.
91. Novak would support this argument, at least according to Newman's reading of him. See Newman (1998):166–167, referring to Novak (1979).
92. CCAR (1950).
93. Kravitz (2006):93.
94. Glick (1999):50.
95. BT *Arachin* 6b. See Steinberg (2006):338; Sherwin (1974):21; Sherwin (1998):95.
96. Steinberg (2006):338.
97. Ibid., 336.
98. BT *Baba Metzia* 59a. See also Novak (1974):86.
99. Novak (1976):106.
100. Ibid.
101. See, for example, the famous instruction about removing a noisy woodchopper from the vicinity of a dying patient (Moses Isserles on SA *Yoreh De'ah* 339.1). Newman (1998) offers an excellent analysis of this and other texts in this regard.
102. The rule is *ein shaliach l'davar 'averah*—literally, there is no emissary for a transgression. See BT *Kiddushin* 42b; *Baba Kama* 79a; MT *Sheluchin v'Shutafin* 1:1; Tur, *Choshen Mishpat* 182:4. See also Crane (2010):340.
103. Ricoeur (1992):162.

CHAPTER 6

1. Bury (2001):264.
2. Cover (1984).
3. Stemberger (1996):60–61.
4. Ibid., 61.
5. Rubenstein, Wimpfheimer and other scholars of Talmudic narratives echo these observations and arguments.
6. Stemberger (1996):62.
7. Kottek (2003):306.
8. It is not insignificant that the judges are glaringly absent in all the stories of Chananya's end. Robert Cover comments on the distance between judges and mechanisms of state violence, as noted in Cohen (2011):269.
9. Childress (1997).
10. Wimpfheimer (2011):22.
11. Rosner (2006):353.
12. Bleich (1997):77, ad loc; Bleich (1983):xvi; Bleich (1989):xv; Bleich (1996):52, 57.
13. Bury (2001).
14. Arras (1991):37. See also Chambers (1999):7.

15. Childress (1997):258—identifies three types: real, hypothetical but realistic, and hypothetical but fantastic.
16. Arras (1991):37.
17. Childress (1997):262.
18. Wimpfheimer (2011):29. See also p. 45.
19. Newman (2007):185.
20. Childress (1997):264.
21. Ibid.
22. Kunda (1990):480.
23. Ibid., 481.
24. Ibid., 482–483.
25. Jonathan Haidt (2001) expands on this kind of distinction with his social intuitionist model of moral judgment. See Crane and Kadane (2008) for a brief Talmudic discussion on this topic. See also Lehrer (2010) for reporting that includes "subtle omissions and unconscious misperceptions."
26. Mertes and Pennings (2011):127.
27. Sharzer (2011); Cohen (2011).
28. This follows ancient Greek logic: similar cases should be evaluated similarly. Dissimilar cases, however, do not necessarily need to be evaluated dissimilarly. "Consistency does not require that a different moral judgment be made." Mertes and Pennings (2011):122.
29. Mertes and Pennings (2011):122.
30. Ibid., 123.
31. Ibid., 124.
32. Kravitz (1995, 2006); Knobel (1995, 2007).
33. Wimpfheimer (2011):29–30.
34. Sharzer (2011):262.
35. Recall that the centurion in the *Kallah* version also contracts with Chananya.
36. BT *Ketubot* 104a.
37. Sharzer (2011):257.
38. Cohen (2011):273.
39. Ibid., 267.
40. Newman (1998):202.
41. See Introduction to Dorff and Crane (2012).
42. Newman (2007):189ff.
43. Heschel (1955):336–337.
44. Bialik (1917):81.
45. Hauerwas (1973):82.
46. Tucker (2006).
47. Siegel (1966):225; Siegel (1977):127.
48. Dorff (2010).
49. M *Avot* 5.22.
50. Cutter (2006):259.

51. Childress (1997):264.
52. See Cover (1984).
53. Bleich (1977):xviii; Bleich (1983):ix; Bleich (1989):xi; Bleich (1995):ix.
54. Sharzer (2011):261.
55. Bar-Sela et al. (1964):40.
56. Sharzer (2011):262.
57. Ibid., 254.

Bibliography

Abraham, Abraham S. 2006. Euthanasia. In *Jewish Medical Ethics 1989–2004*, Volume 2, ed. Mordechai Halperin, David Fink and Shimon Glick, pp 366–375. Jerusalem: The Dr. Falk Schlesinger Institute for Medical-Halachic Research.

Abrams, Judish Z. 1995. *The Women of the Talmud*. Northvale: Jason Aronson Inc.

Adler, Rachel. 1998. *Engendering Judaism: An Inclusive Theology and Ethics*. Philadelphia: The Jewish Publication Society.

Amudsen, Darrel W. and Gary B. Ferngren. 1995. Medical Ethics, History of the Near and Middle East—Ancient Near East. In *Encyclopedia of Bioethics*, 3rd Edition, Volume 3, ed. Stephen G. Post, pp 1659–1664. New York: Macmillan Reference USA.

Ariès, Philippe. 1974. *Western Attitudes Toward Death: From the Middle Ages to the Present*. Translated by Patricia M. Ranum. Baltimore: The Johns Hopkins University Press.

Arras, John D. 1991. Getting Down to Cases: The Revival of Casuistry in Bioethics. *Journal of Medicine and Philosophy*. 16/1:29–51.

Arras, John D. 1997. Nice Story, But So What? Narrative and Justification in Ethics. In *Stories and Their Limits: Narrative Approaches to Bioethics*, ed. Hilde Lindemann Nelson, pp 65–88. New York: Routledge.

Bardfeld, Philip A. 1976. Jewish Medical Ethics. *Reconstructionist*. 42/6:7–12.

Bar-Sela, Ariel, Hebbel E. Hoff and Elias Faris. 1964. Moses Maimonides' Two Treatises on the Regimen of Health. *Transactions of the American Philosophical Society*. 54/4:3–50.

Barilan, Y. Michael. 2003. Revisiting the Problem of Jewish Bioethics: The Case of Terminal Care. *Kennedy Institute of Ethics Journal*. 13/2:141–168.

Basser, Herbert W. 1984. *Midrashic Interpretations of the Song of Moses*. New York: Peter Lang.

Basser, Herbert W. 1990. Hanina's Torah: A Case of Verse Production or of Historical Fact? In *Approaches to Ancient Judaism*, New Series, Volume I, ed. Jacob Neusner, pp 67–82. Atlanta: Scholars Press.

Belser, Julia W. 2011. Reading Talmudic Bodies: Disability, Narrative, and the Gaze in Rabbinic Judaism. In *Disability in Judaism, Christianity, and Islam: Sacred Texts, Historical Traditions, and Social Analysis*, ed. Darla Schumm and Michael Stoltzfus, pp 5–28. New York: Palgrave Macmillan.

Berger, Natalia. (ed.). 1995. *Jews and Medicine: Religion, Culture, Science.* Philadelphia: The Jewish Publication Society.

Berkowitz, Beth A. 2006. *Execution and Invention: Death Penalty Discourse in Early Rabbinic and Christian Cultures.* New York: Oxford University Press.

Bettan, Israel. 1950. Euthanasia. In *Death and Euthanasia in Jewish Law: Essays and Responsa*, ed. Walter Jacob and Moshe Zemer, pp 123–126. Pittsburgh: Freehof Institute of Progressive Halakhah.

Bialik, Hayyim Nahman. 1917. Halachah and Aggadah. In *Revealment and Concealment.* Translated by Leon Simon, pp 45–87. Jerusalem: Ibis Editions.

Bleich, J. David. 1973. Establishing Criteria of Death. *Tradition.* 13/3:90–113.

Bleich, J. David. 1975. Dignity Lies in the Struggle for Life. *Sh'ma.* 103:20–21.

Bleich, J. David. 1976. Karen Ann Quinlan: A Torah Perspective. *Jewish Life.* Winter. Reprinted in *Contemporary Jewish Ethics*, ed. Menachem Marc Kellner, pp 296–307. New York: Sanhedrin Press. 1978.

Bleich, J. David. 1977. *Contemporary Halakhic Problems.* Volume I. New York: Ktav Publishing House, Inc.

Bleich, J. David. 1983. *Contemporary Halakhic Problems.* Volume II. New York: Ktav Publishing House, Inc.

Bleich, J. David. 1985. Is There an Ethic Beyond Halakhah? In *Studies in Jewish Philosophy*, ed. Norbert Samuelson, pp 527–546. Lanham, MD: University Press of America.

Bleich, J. David. 1989. *Contemporary Halakhic Problems.* Volume III. New York: Ktav Publishing House, Inc.

Bleich, J. David. 1995. *Contemporary Halakhic Problems.* Volume IV. New York: Ktav Publishing House, Inc.

Bleich, J. David. 1996a. Treatment of the Terminally Ill. *Tradition.* 30/3:51–58. Reprinted in *Jewish Ethics and the Care of End-of-Life Patients: A Collection of Rabbinical, Bioethical, Philosophical, and Juristic Opinions*, ed. Peter Joel Hurwitz, Jacques Picard and Avraham Steinberg, pp 57–73. Jersey City: Ktav Publishing House, Inc, 2006.

Bleich, J. David. 1996b. Conjoined Twins. *Tradition.* 31/1:92–125.

Bleich, J. David. 2000. The Case of the British Conjoined Twins. *Tradition.* 34/4:61–78.

Blidstein, Gerald J. 1984. Rabbis, Romans, and Martyrdom—Three Views. *Tradition.* 21/3:54–62.

Borowitz, Eugene B. 1991. *Renewing the Covenant: A Theology for the Postmodern Jew.* Philadelphia: The Jewish Publication Society.

Borowitz, Eugene B. 2009. Response to the Questions. *CCAR Journal.* LVI/IV:26–30.

Boyarin, Daniel. 1999. *Dying for God: Martyrdom and the Making of Christianity and Judaism.* Stanford: Stanford University Press.

Breitowitz, Yitzchok. n.d. Physician-Assisted Suicide: A *Halakhic* Approach. *Jewish Law*. http://www.jlaw.com/Articles/suicide.html.

Brodsky, David. 2006. *A Bride without a Blessing: A Study in the Redaction and Content of Massekhet Kallah and its Gemara*. Tübingen: Mohr Siebeck.

Brody, Baruch. 1983. The Use of Halakhic Material in Discussions of Medical Ethics. *The Journal of Medicine and Philosophy*. 8/3:317–328. Reprinted in Brody (2003).

Brody, Baruch. 2003. *Taking Issue: Pluralism and Casuistry in Bioethics*. Washington, D.C.: Georgetown University Press.

Brothwell, Don and Andrew T. Sandison. 1967. *Diseases in Antiquity: A Survey of the Diseases, Injuries and Surgery of Early Populations*. Springfield: Charles C. Thomas, Publishers.

Bury, Mike. 2001. Illness Narratives: Fact or Fiction? *Sociology of Health and Illness*. 23/3:263–285.

Central Conference of American Rabbis Responsa. 1950. Euthanasia. Available at http://data.ccarnet.org/cgi-bin/respdisp.pl?file=78&year=arr. Accessed June 11, 2012.

Central Conference of American Rabbis Responsa. 1969. Allowing a Terminal Patient to Die. Available at http://data.ccarnet.org/cgi-bin/respdisp.pl?file=77&year=arr. Accessed June 11, 2012.

Central Conference of American Rabbis Responsa. 1980. Euthanasia. Available at http://data.ccarnet.org/cgi-bin/respdisp.pl?file=79&year=arr. Accessed June 11, 2012.

Chambers, Tod. 1999. *The Fiction of Bioethics*. New York: Routledge.

Charon, Rita. 2006. *Narrative Medicine: Honoring the Stories of Illness*. New York: Oxford.

Childress, James F. 1997. Narrative(s) versus Norm(s): A Misplaced Debate in Bioethics. In *Stories and Their Limits: Narrative Approaches to Bioethics*, ed. Hilde Lindemann Nelson, pp 252–271. New York: Routledge.

Cohen, Jonathan. 2011. Jewish Bioethics: Between Interpretation and Criticism. In *Midrash and Medicine: Healing Body and Soul in the Jewish Interpretive Tradition*, ed. William Cutter, pp 263–273. Woodstock: Jewish Lights Publishing.

Cover, Robert. 1984. *Nomos* and Narrative. *Harvard Law Review*. 97/4:4–68.

Cover, Robert. 1995. *Narrative, Violence, and the Law: The Essays of Robert Cover*, ed. Martha Minow, Michael Ryan and Austin Sarat. Ann Arbor, MI: University of Michigan Press.

Crane, Jonathan K. 2008. With a Mighty Hand: A Judaic Perspective on the Ethics of Modern Armed Conflict. In *Enemy Combatants, Terrorism, and Armed Conflict Law: A Guide to the Issues*, Praeger Security International series. ed. David Linnan, pp 184–206. Westport: Praeger.

Crane, Jonathan K. 2009a. *Rhetoric of Modern Jewish Ethics*. PhD Dissertation, Unpublished. Toronto: University of Toronto.

Crane, Jonathan K. 2009b. *Naaseh V'Nishma*: For Rabbi Eugene B. Borowitz. *CCAR Journal*. LVI/IV:21–25.

Crane, Jonathan K. 2010. Defining the Unspeakable: Incitement in Halakhah and Anglo-American Jurisprudence. *Journal of Law and Religion*. XXV/2:329–356.

Crane, Jonathan K. 2011a. Torture: Judaic Twists. *Journal of Law and Religion*. XXVI/2:469–504.

Crane, Jonathan K. 2011b. Shameful Ambivalences: Dimensions of Rabbinic Shame. *AJS Review: The Journal of the Association for Jewish Studies*. 35/1:61–84.

Crane, Jonathan K. 2011c. Torturous Ambivalence: Judaic Struggles with Torture. *Journal of Religious Ethics*. 39/4:598–605.

Crane, Jonathan K. 2012. Rethinking Conjoined Twins. *CCAR Journal*. LIX/I:125–141.

Crane, Jonathan K. Medical Futility and Liturgical Efficacy: When Clinicians Pray for Patients to Die. Unpublished.

Crane, Jonathan K. and Joseph B. Kadane. 2008. Seeing Things: The Internet, The Talmud and Anais Nin. *Review of Rabbinic Judaism*. 11/2:342–344.

Cutter, William. 1995. Rabbi Judah's Handmaid: Narrative Influence on Life's Important Decisions. In *Death and Euthanasia in Jewish Law: Essays and Responsa*, ed. Walter Jacob and Moshe Zemer, pp 61–87. Pittsburgh: Freehof Institute of Progressive Halakhah.

Cutter, William. 2006. Do the Qualities of Story Influence the Quality of Life? Some Perspectives on the Limitations and Enhancements of Narrative Ethics. In *Quality of Life in Jewish Bioethics*, ed. Noam Zohar, pp 55–66. Lanham, MD: Lexington Books.

Cutter, William. (ed.). 2007. *Healing and the Jewish Imagination: Spiritual and Practical Perspectives on Judaism and Health*. Woodstock, VT: Jewish Lights Publishing.

Cutter, William. (ed.). 2011. *Midrash and Medicine: Healing Body and Soul in the Jewish Interpretive Tradition*. Woodstock, VT: Jewish Lights Publishing.

Dagi, Teodoro Forcht. 1975. The Paradox of Euthanasia. *Judaism*. 24/2:157–167.

Dorff, Elliot N. 1998. *Matters of Life and Death: A Jewish Approach to Modern Medical Ethics*. Philadelphia: The Jewish Publication Society.

Dorff, Elliot N. 1999. Assisted Suicide. *Journal of Law and Religion*. 13/2:263–287.

Dorff, Elliot N. 2000. End-Stage Medical Care: Halakhic Concepts and Values. In *Life and Death Responsibilities in Jewish Biomedical Ethics*, ed. Aaron L. Mackler, pp 309–337. New York: the Jewish Theological Seminary of America.

Dorff, Elliot N. 2006. These and Those Are the Words of the Living God: Talmudic Sound and Fury in Shaping National Policy. In *Handbook of Bioethics and Religion*, ed. David E. Guinn, pp 143–168. New York: Oxford University Press.

Dorff, Elliot N. 2010. Applying Jewish Law to New Circumstances. In *Tiferet Leyisrael: Jubilee Volume in Honor of Israel Francus*, ed. Joel Roth, Menahem Schmelzer and Yaacov Francus, pp 189–199. New York: Jewish Theological Seminary.

Dorff, Elliot N. and Jonathan K. Crane (eds.). 2012. *The Oxford Handbook of Jewish Ethics and Morality*. New York: Oxford University Press.

Droge, Arthur J. and James D. Tabor. 1992. *A Noble Death: Suicide and Martyrdom Among Christians and Jews in Antiquity*. San Francisco: HarperSanFrancisco.

Elon, Menachem. 1994. *Jewish Law: History, Sources, Principles*. Translated by Bernard Auerbach and Melvin J. Sykes. Philadelphia: Jewish Publication Society.

Epstein, Louis M. 1935. The Institution of Concubinage among the Jews. *Proceedings of the American Academy for Jewish Research*. VI:153–188.

Falk, Von Ze'ev W. 1996. Euthanasia and Judaism. *Zeitschrift für Evangelische Ethik*. 40:170–174.

Falk, Von Ze'ev W. 1998. Jewish Perspectives on Assisted Suicide and Euthanasia. *Journal of Law and Religion*. 13/2:379–384.

Finkelstein, Louis. 1938. The Ten Martyrs. In *Essays and Studies in Memory of Linda R. Miller*, ed. Israel Davidson, pp 29–55. New York: The Jewish Theological Seminary of America.

Fox, Marvin (ed.). 1975. *Modern Jewish Ethics: Theory and Practice*. Cincinnati: Ohio State University Press.

Freedman, Benjamin. 1999. *Duty and Healing: Foundations of a Jewish Bioethic*. New York: Routledge.

Friedenwald, Harry. 1944. *The Jews and Medicine*. 2 vols. Baltimore: Johns Hopkins University Press.

Freund, Richard A. 1990. *Understanding Jewish Ethics*. San Francisco: Edwin Mellin Text.

Gaster, Moses. 1968. *The Exempla of the Rabbis*. New York: Ktav Publishing House, Inc.

Gazzaniga, Michael. 1998. *The Mind's Past*. Berkeley: University of California Press.

Gesundheit, Benjamin, Avraham Steinberg, Shimon Glick, Reuven Or and Alan Jotkowitz. 2006. Euthanasia: An Overview and the Jewish Perspective. *Cancer Investigation*. 24:621–629.

Gibbs, Robert. 2004. *Gesetz* in *The Star of Redemption*. In *Rosenzweig als Leser. Kontextuelle Kommentare zum "Stern der Erlösung"*, ed. Max Niemeyer von Martin Brasser, pp 395–410. Tübingen: Max Niemeyer Verlag.

Ginzberg, Louis. 1936. *The Legends of the Jews*. Philadelphia: Jewish Publication Society.

Glick, Shimon. 1999. The Jewish Approach to Living and Dying. In *Jewish and Catholic Bioethics: An Ecumenical Dialogue*, ed. Edmund D. Pellegrino and Alan I. Faden, pp 43–53. Washington, D.C.: Georgetown University Press.

Goldberg, Michael. 1984. The Story of the Moral: Gifts or Bribes in Deuteronomy? *Interpretations.* 38/1:15–25.

Goldberg, Michael. 1991. *Jews and Christians: Getting Our Stories Straight.* Philadelphia, PA: Trinity Press International.

Goldberg, Zalman Nehemiah. 2006. Approaching Death: Synopsis of the Responsa. In *Jewish Medical Ethics 1989–2004*, Volume 2, ed. Mordechai Halperin, David Fink and Shimon Glick, pp 344–346. Jerusalem: The Dr. Falk Schlesinger Institute for Medical-Halachic Research.

Goldstein, Sidney. 1989. *Suicide in Rabbinic Literature.* Hoboken: Ktav Publishing House, Inc.

Goodman, Lenn. 2008. *Love Thy Neighbor as Thyself.* New York: Oxford University Press.

Gordis, David. 1989. Wanted—The Ethical in Jewish Bio-Ethics. *Judaism.* 38/1:28–40.

Gordon, Benjamin Lee. 1949. *Medicine throughout Antiquity.* Philadelphia: F. A. Davis Company, Publishers.

Gray, Hillel C. 2009. *Foreign Features in Jewish Law: How Christian and Secular Moral Discourses Permeate Halakhah.* Chicago: University of Chicago. PhD Dissertation.

Green, Ronald M. 1999. Jewish Teaching on the Sanctity and Quality of Life. In *Jewish and Catholic Bioethics: An Ecumenical Dialogue*, ed. Edmund D. Pellegrino and Alan I. Faden, pp 25–42. Washington, D.C.: Georgetown University Press.

Greenwood, Daniel J. H. 1997. Akhnai: Legal Responsibility in the World of the Silent God. *Utah Law Review.* 309–358. Available at http://papers.ssrn.com/sol3/papers.cfm?abstract_id=794784. Accessed February 26, 2012.

Gross, Abraham. 2005. *Spirituality and Law: Courting Martyrdom in Christianity and Judaism.* Lanham: University Press of America, Inc.

Haidt, Jonathan. 2001. The Emotional Dog and Its Rational Tail: A Social Intuitionist Approach to Moral Judgment. *Psychological Review.* 108/4:814–834.

Hammer, Reuven. 1986. *Sifre: A Tannaitic Commentary on the Book of Deuteronomy.* New York: Yale University Press.

Harvey, Warren Zev. 2007. Aggadah in Maimonides' *Mishneh Torah. Diné Israel.* 24:197–207.

Hauerwas, Stanley. 1973. The Self as Story: Religion and Morality from the Agent's Perspective. *Journal of Religious Ethics.* 1:73–85.

Hauerwas, Stanley. 1993. Casuistry as a Narrative Art. *Interpretation.* 37:377–388.

Hauerwas, Stanley and L. Gregory Jones (eds.). 1997. *Why Narrative? Readings in Narrative Theology.* Eugene, OR: Wipf and Stock Publishers.

Hayes, Christine E. 1995. Amoraic Interpretations and Halakhic Development: The Case of the Prohibited Basilica. *Journal for the Study of Judaism in the Persian, Hellenistic and Roman Period.* 26/2:156–168.

Herring, Basil F. 1984. *Jewish Ethics and Halakhah for Our Time: Sources and Commentary*. New York: Ktav Publishing House, Inc.

Heschel, Abraham J. 1955. *God in Search of Man: A Philosophy of Judaism*. New York: Farrar, Strauss, and Giroux.

Hochhoizer, Rivka. 1992. *Chayim—b'chol mechir? ba-machashavah hayehudit uvafilosofyah ba'et hachadashah* [Life at All Costs? Or Euthanasia and Medical Ethics in Jewish Thought and in Philosophy]. Shefar'am: Almashrak. [Hebrew].

Idel, Moshe. 1990. *Golem: Jewish Magical and Mystical Traditions on the Artificial Anthropoid*. Albany: Sate University of New York Press.

Jacob, Walter. 1985. Quality of Life and Euthanasia. In *Central Conference of American Rabbis Responsa*. Available at http://www.ccarnet.org/responsa/carr-138-140/. Accessed April 20, 2012.

Jacob, Walter. 1995. End-State Euthanasia. In *Death and Euthanasia in Jewish Law*, ed. Walter Jacob and Moshe Zemer, pp 89–103. Pittsburgh: Freehof Institute of Progressive Halakhah.

Jastrow, Marcus. 1903. *A Dictionary of the Targumim, the Talmud Babli and Yerushalmi, and the Midrashic Literature with an Index of Scriptural Quotations*. New York: G. P. Putnam.

Jonson, Albert R. and Stephen Toulmin. 1988. *The Abuse of Casuistry: A History of Moral Reasoning*. Berkeley: University of California Press.

Jotkowitz, Alan. 2012. The Use of Narrative in Jewish Medical Ethics. *Journal of Religion and Health*. Available at http://dx.doi.org/10.1007/s10943-012-9585-x. Accessed April 20, 2012.

Jotkowitz, Alan, and S. Glick. 2009. Navigating the Chasm between Religious and Secular Perspectives in Modern Bioethics. *Journal of Medical Ethics*. 35:357–360.

Kanarek, Jane. 2010. He Took the Knife: Biblical Narrative and the Formation of Rabbinic Law. *AJS Review*. 34:1:65–90.

Katzoff, Ranon. 1993. Roman Edicts and *Ta'anit* 29a. *Classical Philology*. 88/2:141–144.

Kierkegaard, Søren A. 1959. *The Journals of Kierkegaard*. Translated and edited by Alexander Dru. New York: Harper Torchbooks.

Knobel, Peter. 1995. Suicide, Assisted Suicide, Active Euthanasia. In *Death and Euthanasia in Jewish Law*, ed. Walter Jacob and Moshe Zemer, pp 27–59. Pittsburgh: Freehof Institute of Progressive Halakhah.

Knobel, Peter. 2007. An Expanded Approach to Jewish Bioethics: A Liberal/Aggadic Approach. In *Healing and the Jewish Imagination: Spiritual and Practical Perspectives on Judaism and Health*, ed. W. Cutter, pp 171–182. Woodstock: Jewish Lights Publishing.

Kottek, Samuel. 2003. Medicine in Ancient Hebrew and Jewish Cultures. In *Medicine across Cultures: History and Practice of Medicine in Non-Western Cultures*, ed. Helaine Selin, pp 305–324. Dordrecht: Kluwer Academic Publishers.

Kunda, Ziva. 1990. The Case for Motivated Reasoning. *Psychological Bulletin*. 108/3:480–498.

Kravitz, Leonard S. 1995. Euthanasia. In *Death and Euthanasia in Jewish Law*, ed. Walter Jacob and Moshe Zemer, pp 11–25. Pittsburgh: Freehof Institute of Progressive Halakhah.

Kravitz, Leonard S. 2006. "Some" Reflections on Jewish Tradition and the End-of-Life Patient. In *Jewish Ethics and the Care of End-of-Life Patients*, ed. Peter Joel Hurwitz, Jacques Picard and Avraham Steinberg, pp 75–97. Jersey City: Ktav Publishing House, Inc.

Lehrer, Jonah. December 13, 2010. The Truth Wears Off: Is There Something Wrong with the Scientific Method? *The New Yorker*. 52ff.

Leiman, Sid Z. 1977. The Karen Ann Quinlin Case: A Jewish Perspective. *Gratz College Annual of Jewish Studies*. VI:43–50.

Levine, Bob. 1975. The Terminally Ill: May We Let Them Die? *Sh'ma*. 103:17–18.

Lichtenstein, Aaron. 1975. Does Judaism Recognize an Ethic Independent of Halakhah? In *Modern Jewish Ethics: Theory and Practice*, ed. Marvin Fox, pp. 62–88. Columbus: Ohio State University Press.

Lieberman, Saul. 1975. Redifat Dat Yisrael [Persecution of Jewish Religion]. In *Salo Wittmayer Baron Jubilee Volume on the Occasion of His Eightieth Birthday*, ed. Saul Lieberman and Arthur Hyman, pp 213–245. Jerusalem: American Academy for Jewish Research. [Hebrew].

Lifshitz, Berachyahu. 2007. Aggadah Versus Haggadah: Towards a More Precise Understanding of the Distinction. *Diné Israel*. 24:11–28.

Lorberbaum, Yair. 2007. Reflections on the Halakhic Status of Aggadah. *Diné Israel*. 24:29–64.

Luban, David. 2004. The Coiled Serpent of Argument: Reason, Authority, and Law in a Talmudic Tale. *Chicago-Kent Law Review*. 79:1253–1288.

MacIntyre, Alasdair. 1984. *After Virtue*. 2nd Edition. Notre Dame: University of Notre Dame Press.

Mackler, Aaron L. 2003. *Introduction to Jewish and Catholic Bioethics: A Comparative Analysis*. Washington, D.C.: Georgetown University Press.

Marquez, Gabriel Garcia. 1982. *Chronicle of a Death Foretold*. New York: Random House.

Nelson, Hilde Lindemann (ed.). 1997. *Stories and Their Limits: Narrative Approaches to Bioethics*. New York: Routledge.

Newman, Louis E. 1998. *Past Imperatives: Studies in the History and Theory of Jewish Ethics*. Albany: State University of New York.

Newman, Louis E. 2005. *An Introduction to Jewish Ethics*. Upper Saddle River, NJ: Pearson Prentice Hall.

Newman, Louis E. 2007. The Narrative and the Normative: The Value of Stories for Jewish Ethics. In *Healing and Jewish Imagination: Spiritual and Practical Perspectives on Judaism and Health*, ed. William Cutter, pp 183–192. Woodstock, VT: Jewish Lights Publishing.

Novak, David. 1979. Judaism and Contemporary Bioethics. *The Journal of Medicine and Philosophy*. 4/4:347–366.

Novak, David. 1983. *The Image of the Non-Jew in Judaism: An Historical and Constructive Study of the Noahide Laws*. Lewiston: The Edwin Mellen Press.

Novak, David. 2005. *The Jewish Social Contract: An Essay in Political Theory*. Princeton: Princeton University Press.

Novak, David. 2007. *The Sanctity of Human Life*. Washington, D.C.: Georgetown University Press.

O'Mathúna, Dónal P. and Darrel W. Amundsen. 1998. Historical and Biblical References in Physician-Assisted Suicide Court Opinions. *Notre Dame Journal of Law, Ethics and Public Policy*. 12:473–496.

Preuss, Julius. 1971. *Biblish-talmudische Medizin: Beiträge zur Geschichte der Heilkunde und der Kulture überhaupt*. Introduction by Sussmann Muntner. New York: Ktav Publishing House, Inc.

Preuss, Julius. 1993. *Biblical and Talmudic Medicine*. Translated and edited by Fred Rosner. Northvale: Jason Aronson Inc.

Prouser, Joseph H. 1997. Being of Sound Mind and Judgment: Rethinking Sanctions in the Case of Assisted Suicide. *Conservative Judaism*. 49/4:3–16.

Rackman, Emanuel. 1956. Morality in Medico-Legal Problems: A Jewish View. *New York University Law Review*. 31:1205–1214.

Raskas, Bernard S. 1975. When a Life Is No Life—The Right to Die. *Sh'ma*. 103:18–19.

Reconstructionist Rabbinical College Center for Jewish Ethics. 2002. *Behoref Hayamim/In the Winter of Life: A Values-based Jewish Guide for Decision-making at the End of Life*. Wyncote: Reconstructionist Rabbinical College Press.

Reisner, Avram I. 2000. Care for the Terminally Ill: Halakhic Concepts and Values. In *Life and Death Responsibilities in Jewish Biomedical Ethics*, ed. Aaron L. Mackler, pp 239–264. New York: The Jewish Theological Seminary of America.

Resnicoff, Steven H. 1998. Physician-Assisted Suicide under Jewish Law. *Jewish Law*. http://www.jlaw.com/Articles/phys-suicide.html.

Resnicoff, Steven H. 2003. Euthanasia and Physician-Assisted Suicide in Jewish Law. In *Public Policy and Social Issues: Jewish Sources and Perspectives*, ed. Marshall J. Breger, pp 79–107. Wesport: Praeger.

Ricoeur, Paul. 1992. *Oneself as Another*. Translated by Kathleen Blamey. Chicago: University of Chicago Press.

Ricoeur, Paul. 1995. *Figuring the Sacred: Religion, Narrative, and Imagination*. Translated by David Pellauer. Minneapolis: Fortress Press.

Rosner, Fred. 1967. The Jewish Attitude toward Euthanasia. *New York State Journal of Medicine* 67 (September). Reprinted in *Jewish Bioethics*, ed. Fred Rosner and J. David Bleich, pp 253–265. New York: Sanhedrin Press.

Rosner, Fred. 1970. Suicide in Jewish Law. *Tradition* (Summer). Reprinted in *Jewish Bioethics*, ed. Fred Rosner and J. David Bleich, pp 317–330. New York: Sanhedrin Press.

Rosner, Fred. 1971. Contraception in Jewish Law. *Tradition* 13/1 (Fall):90–103. Reprinted in *Jewish Bioethics*, ed. Fred Rosner and J. David Bleich, pp 86–96. New York: Sanhedrin Press.

Rosner, Fred. 1986. Risks Versus Benefits in Treating the Gravely Ill Patient: Ethical and Religious Considerations. In *Jewish Values in Bioethics*, ed. Levi Meier, pp 46–56. New York: Human Sciences Press, Inc.

Rosner, Fred. 1988. Rabbi Moshe Feinstein on the Treatment of the Terminally Ill. *Judaism*. 37/2:188–198.

Rosner, Fred. 1995a. Euthanasia. In *Contemporary Jewish Ethics and Morality*, ed. Elliot N. Dorff and Louis E. Newman, pp 350–362. New York: Oxford University Press.

Rosner, Fred. 1995b. *Medicine in the Bible and the Talmud: Selections from Classical Jewish Sources*. New York: Ktav Publishing House, Inc.

Rosner, Fred. 1998. Suicide in Jewish Law. In *Jewish Approaches to Suicide, Martyrdom, and Euthanasia*, ed. Kalman J. Kaplan and Matthew B. Schwartz, pp 61–77. Northvale: Jason Aronson Inc.

Rosner, Fred. 2006. Jewish Perspectives on Death and Dying. In *Jewish Medical Ethics 1989–2004*, Volume 2, ed. Mordechai Halperin, David Fink and Shimon Glick, pp 347–365. Jerusalem: The Dr. Falk Schlesinger Institute for Medical-Halachic Research.

Sharzer, Leonard A. 2001. Artificial Hydration and Nutrition: Revisiting the Dorff and Reisner *Teshuvot*. *Conservative Judaism*. 53/2:60–68.

Sharzer, Leonard A. 2011. Aggadah and Midrash: A New Direction for Bioethics? In *Midrash and Medicine*, ed. W. Cutter, pp 245–262. Woodstock: Jewish Lights Publishing.

Sherwin, Byron L. 1974. Jewish Views on Euthanasia. *The Humanist*. 34/4:19–21.

Sherwin, Byron L. 1987. Theodicy. In *Contemporary Jewish Religious Thought*, ed. Arthur A. Cohen and Paul Mendes-Flohr, pp 959–970. New York: The Free Press.

Sherwin, Byron L. 1995. A View of Euthanasia. In *Contemporary Jewish Ethics and Morality*, ed. Elliot N. Dorff and Louis E. Newman, pp 363–381. New York: Oxford University Press.

Sherwin, Byron L. 1998. Euthanasia as a Halakhic Option. In *Jewish Approaches to Suicide, Martyrdom, and Euthanasia*, ed. Kalman J. Kaplan and Matthew B. Schwartz, pp 80–97. Northvale: Jason Aronson Inc.

Shulman, Nisson E. 1998. *Jewish Answers to Medical Ethics Questions*. Northvale: Jason Aronson Inc.

Siegel, Seymour. 1966. *The Condition of Jewish Belief: A Symposium, Compiled by the Editors of Commentary Magazine*. New York: Macmillan.

Siegel, Seymour. 1977. *Conservative Judaism and Jewish Law*. New York: The Rabbinical Assembly.

Silverman, Morris. 1986. *High Holidays Prayer Book*. Bridgeport: The Prayer Book Press.

Simon-Shoshan, Moshe. 2007. Halakhic Mimesis: Rhetorical and Redactional Strategies in Tannaitic Narrative. *Diné Israel*. 24:101–123.

Smith, Morton (trans.). 2009. *Hekhalot Rabbati*: The Greater Treatise Concerning the Palaces of Heaven. Corrected by Gershom Scholem. Transcribed and edited by Don Karr. http://www.digital-brilliance.com/contributed/Karr/HekRab/HekRab.pdf. Accessed February 24, 2012.

Steinberg, Avraham. 1999. The Meaning of Suffering: A Judaic Perspective, In *Jewish and Catholic Bioethics: An Ecumenical Dialogue*, ed. Edmund D. Pellegrino and Alan I. Faden, pp 77–82. Washington, D.C.: Georgetown University Press.

Steinberg, Avraham. 2004. Jewish Medical Ethics. In *Jewish Medical Ethics*, Volume I, ed. Mordechai Halperin, David Fink and Shimon Glick, pp 28–42. Jerusalem: The Dr. Falk Schlesinger Institute for Medical-Halakhic Research.

Steinberg, Avraham. 2006. The Terminally Ill Patient. In *Jewish Medical Ethics 1989–2004*, Volume 2, ed. Mordechai Halperin, David Fink and Shimon Glick, pp 321–342. Jerusalem: The Dr. Falk Schlesinger Institute for Medical-Halachic Research.

Stemberger, Günter. 1996. *Introduction to the Talmud and Midrash*. Second Edition. Trans. and ed., Markus Bockmuehl. Edinburgh: T&T Clark.

Stone, Suzanne Last. 1993. In Pursuit of the Counter-Text: The Turn to the Jewish Legal Model in Contemporary American Legal Theory. *Harvard Law Review*. 106/4:813–894.

Stone, Suzanne Last. 2007. On the Interplay of Rules, "Cases," and Concepts in Rabbinic Legal Literature: Another Look at the Aggadot on Honi the Circle-Drawer. *Diné Israel*. 24:125–155.

Sussman, Max. 1967. Diseases in the Bible and the Talmud. In *Diseases in Antiquity: A Survey of the Diseases, Injuries and Surgery of Early Populations*, ed. Don Brothwell and A. T. Sandison, pp 209–221. Springfield: Charles C. Thomas, Publisher.

Telushkin, Joseph. 2009. *A Code of Jewish Ethics*. Volume 2: Love Your Neighbor as Yourself. New York: Bell Tower.

Tendler, Moshe D. 1996. *Responsa of Rav Moshe Feinstein: Translation and Commentary*. New York: Ktav Publishing House, Inc.

Tigay, Jeffrey H. 1996. *The JPS Torah Commentary: Deuteronomy*. Philadelphia: The Jewish Publication Society.

Tucker, Gordon. 2006. Halakhic and Metahalakhic Arguments Concerning Judaism and Homosexuality. Submitted to the Conservative Movement's Committee on Jewish Law and Standards. Available at http://www.rabbinicalassembly.org/sites/default/files/public/halakhah/teshuvot/20052010/tucker_homosexuality.pdf. Accessed July 3, 2012.

van Henten, Jan Willem. 2004. Jewish and Christian Martyrs. In *Saints and Role Models in Judaism and Christianity*, ed. Marcel Poorthuis and Joshua Schwartz, pp 163–182. Leiden: Brill.

Waldenberg, Eliezer Yehudah. *Responsa Tzitz Eliezer*, 5, Ramat Rachel, §28.

Walker, Mary Jean. 2012. *Neuroscience, Self-Understanding and Narrative Truth. American Journal of Bioethics Neuroscience*. 3/4:63–74.

Weiner, Yaakov. 1995. *Ye Shall Surely Heal: Medical Ethics from a Halakhic Perspective*. Jerusalem: Jerusalem Center for Research.

Wimpfheimer, Barry S. 2004. "But It Is Not So": Toward a Poetics of Legal Narrative in the Talmud. *Prooftexts*. 24:51–86.

Wimpfheimer, Barry S. 2007. Talmudic Legal Narrative: Broadening the Discourse of Jewish Law. *Diné Israel*. 24:157–196.

Wimpfheimer, Barry S. 2011. *Narrating the Law: A Poetics of Talmudic Legal Stories*. Philadelphia: University of Pennsylvania Press.

Wolpe, Paul Root. 2002. Ending Life. In *Behoref Hayamim/In the Winter of Life: A Values-Based Jewish Guide for Decision-Making at the End of Life*, pp 134–147. Wyncote: Reconstructionist Rabbinical College Press.

Yuter, Alan J. 2001. *Hora'at Sha'ah*: The Emergency Principle in Jewish Law and a Contemporary Application. *Jewish Political Studies Review*. 13/3–4. Available at: http://jcpa.org/jpsr/jpsr-yuter-f01.htm. Accessed June 1, 2012.

Zlotnick, Dov. 1966. *The Tractate "Mourning" (Semahot): Regulations Relating to Death, Burial, and Mourning*. New Haven: Yale University Press.

Zohar, Noam J., (ed.). 2006. *Quality of Life in Jewish Bioethics*. Lanham: Lexington Books.

Zoloth, Laurie. 1999. *Health Care and the Ethics of Encounter: A Jewish Discussion of Social Ethics*. Chapel Hill: University of North Carolina Press.

INDEX

Notes: Locators followed by 'n' refer to note numbers.

Source Index

Notes: Locators followed by 'n' refer to note numbers.

Manuscripts and Variations of the Chananya Story

Other Rabbinic Sources